SUSTAINABILITY IN INDUSTRY 4.0

Mathematical Engineering, Manufacturing, and Management Sciences

Series Editor: Mangey Ram, Professor, Assistant Dean (International Affairs), Department of Mathematics, Graphic Era University, Dehradun, India

The aim of this new book series is to publish the research studies and articles that bring up the latest development and research applied to mathematics and its applications in the manufacturing and management sciences areas. Mathematical tool and techniques are the strength of engineering sciences. They form the common foundation of all novel disciplines as engineering evolves and develops. The series will include a comprehensive range of applied mathematics and its application in engineering areas such as optimization techniques, mathematical modelling and simulation, stochastic processes and systems engineering, safety-critical system performance, system safety, system security, high assurance software architecture and design, mathematical modelling in environmental safety sciences, finite element methods, differential equations, reliability engineering, etc.

Non-Linear Programming
A Basic Introduction
Nita H. Shah and Poonam Prakash Mishra

Applied Soft Computing and Embedded System Applications in Solar Energy
Rupendra Kumar Pachauri, J. K. Pandey, Abhishek Sharma, Om Prakash Nautiyal, Mange Ram

Differential Equations in Engineering
Research and Applications
Edited by Nupur Goyal, Piotr Kulczycki, and Mangey Ram

Sustainability in Industry 4.0
Challenges and Remedies
Edited by Shwetank Avikal, Amit Raj Singh, Mangey Ram

Applied Mathematical Modeling and Analysis in Renewable Energy
Edited by Manoj Sahni and Ritu Sahni

For more information about this series, please visit: https://www.routledge.com/Mathematical-Engineering-Manufacturing-and-Management-Sciences/book-series/CRCMEMMS

SUSTAINABILITY IN INDUSTRY 4.0
Challenges and Remedies

Edited by
Shwetank Avikal, Amit Raj Singh, and Mangey Ram

CRC Press
Taylor & Francis Group
Boca Raton London New York

CRC Press is an imprint of the
Taylor & Francis Group, an **Informa** business

First edition published 2022
by CRC Press
6000 Broken Sound Parkway NW, Suite 300, Boca Raton, FL 33487-2742

and by CRC Press
2 Park Square, Milton Park, Abingdon, Oxon, OX14 4RN

Library of Congress Cataloging-in-Publication Data
Names: Avikal, Shwetank, editor. | Singh, Amit Raj, editor. | Ram, Mangey, editor.
Title: Sustainability in industry 4.0 : challenges and remedies / edited by
Shwetank Avikal, Amit Raj Singh, and Mangey Ram.
Description: First edition. | Boca Raton : CRC Press, 2022 |
Series: Mathematical engineering, manufacturing, and management sciences | Includes
bibliographical references and index.
Identifiers: LCCN 2021020824 (print) | LCCN 2021020825 (ebook) |
ISBN 9780367607739 (hbk) | ISBN 9780367608712 (pbk) |
ISBN 9781003102304 (ebk)
Subjects: LCSH: Industry 4.0. | Sustainable engineering. | Manufacturing processes.
Classification: LCC T59.6 .S87 2022 (print) | LCC T59.6 (ebook) | DDC 666/.14dc23
LC record available at https://lccn.loc.gov/2021020824
LC ebook record available at https://lccn.loc.gov/2021020825

ISBN: 978-0-367-60773-9 (hbk)
ISBN: 978-0-367-60871-2 (pbk)
ISBN: 978-1-003-10230-4 (ebk)

DOI: 10.1201/9781003102304

Typeset in Times
by MPS Limited, Dehradun

Contents

Preface

Industry 4.0 (the fourth industrial revolution) lays the foundation of future industries incorporating digitalization and the intelligentization of the manufacturing process. Industry 4.0 facilitates the industries for a paradigm shift from mass production to customized production scenario enabling better control of the value chain of the products on real-time engagement. A large and growing number of manufacturers are realizing substantial financial and environmental benefits from sustainable business practices. Sustainable manufacturing is the creation of manufactured products through economically-sound processes that minimize negative environmental impacts while conserving energy and natural resources. Sustainable manufacturing also enhances employee, community and product safety.

The aim of this book is to provide a floor for researchers, industries, managers and government policy makers to cooperate and collaborate among themselves to develop more sustainable societies, industries need to better understand how to respond to environmental, economic and social challenges and transform industrial behaviour. The objective of this book is to provide knowledge and accelerate the transition towards a sustainable industrial system.

The edited book "*Sustainability in Industry 4.0: Challenges and Remedies*" gripped on a comprehensive range of techniques and theories of engineering and management applied in the field of Industrial Management. Various mathematical tools, techniques, strategy and methods in engineering management applications will be discussed in the chapters, which is the unique point of this book.

This edited issue of Mathematical Engineering, Manufacturing, and Management Sciences "*Sustainability in Industry 4.0: Challenges and Remedies*" includes invited papers appropriate to the theme and the complex solution approaches to handle the research challenges in the related domain. The topics covered are organized as follows:

Chapter 1 provides insights into to the possible solutions provided by Industry 4.0 for achieving these goals. This chapter also provides a bird's eye view of various possibilities and constraints for using Industry 4.0 as a tool for sustainability.

Chapter 2 discusses about the Sustainability of large-scale industries in global market. This chapter discusses the role of large-scale industries with regards to conceptual, operational and futuristic positioning of sustainability in the global market.

Chapter 3 discusses about the Sustainable supplier selection in industry 4.0. In this work, a three-stage robust decision frame work for sustainable supplier selection based on the Fuzzy Kano model (FKM) and Fuzzy Inference System (FIS) has been proposed.

Chapter 4 discusses about the Supply chain and the sustainability management in Industry 4.0. This chapter presents the supply chain's performance management and optimization by suggesting a sustainable supplier performance method for the manufacturing industry.

Chapter 5 represents a WASPAS based multi-criteria decision-making approach for selecting oxygen delignification additives in pulp and paper industry.

Chapter 6 discusses a case study that given insight into Indian consumers' behaviour towards green product consumption.

Chapter 7 discusses about the Energy consumption optimization based on six sigma tool (DAMIC) and energy value stream mapping in Industry.

Chapter 8 provides a study that aims to identify the significant challenges that may emerge while applying Machine Learning techniques in the Additive Manufacturing (AM) processes.

Chapter 9 provides a review on sustainability performance parameters, its measurement methods, and challenges.

Chapter 10 presents a literature and highlight the current status of research on industrial sustainability. The finding of the study shows empirical methodology, qualitative technique and analysis at country level specifically in developing countries.

Chapter 11 discusses about the An AHP based approach to determine the Effects of COVID-19 on Industrial Sustainability.

Chapter 12 discusses about the Roadmap of Smart Manufacturing for Developing Countries. The study concludes that Industry-4.0 is future for manufacturing and must to withstand in global competition.

Chapter 13 explain the concept of Sustainable Entrepreneurship and how the sustainability of the business is beneficial for small and medium-sized enterprises.

The book shall be a valuable tool for practitioners and managers of Industry.

The editors are grateful to various authors across the globe who have contributed for this editorial work. All the reviewers who have helped through their comments and suggestions in improving the quality of the chapters deserve significant praise for their assistance.

Finally and most importantly, all the editors Dr Shwetank Avikal, Dr. Amit Raj Singh and Prof. Mangey Ram would like to dedicate this editorial to his family members and friends.

Shwetank Avikal
Graphic Era Hill University Dehradun, India
Amit Raj Singh
National Institute of Technology Raipur, India
Mangey Ram
Graphic Era Deemed to be University, India

Acknowledgements

The Editors acknowledges CRC press for this opportunity and professional support. Also, we would like to thank all the chapter authors and reviewers for their availability for this work.

Editor's Biographies

Dr. Shwetank Avikal has completed his M.Tech from Motilal Nehru National Institute of Technology (MNNIT-ALD) Allahabad, India in 2009 in Mechanical Engineering. He has also completed his Ph.D. in Mechanical Engineering from the same (MNNIT-ALD) University in 2016. Now a days, he is working as Assistant Professor at Mechanical Engineering Department, Graphic Era Hill University, Dehradun, India since 2013. His area of Interest is production and industrial Engineering, Operations Research, Decision Science, Soft Computing, Mathematical Modelling, Sustainability, Line Balancing etc. He has published over 40 Research articles in referred International Journals and Conferences.

Dr. Amit Raj Singh has completed his M.Tech from Motilal Nehru National Institute of Technology (MNNIT-ALD) Allahabad, India in 2007 in Mechanical Engineering. He has also completed his Ph.D. in Mechanical Engineering from the same (MNNIT-ALD) University in 2012. Now a days, he is working as Assistant Professor at Mechanical Engineering Department, NIT Raipur, India since 2012. His area of Interest is production and industrial Engineering, Operations Research, Decision Science, Soft Computing, Mathematical Modelling, Supply Chain Management, Sustainability etc. He has published over 50 Research articles in referred International Journals and Conferences.

Dr. Mangey Ram received the Ph.D. degree major in Mathematics and minor in Computer Science from G. B. Pant University of Agriculture and Technology, Pantnagar, India, in 2008. He has been a Faculty Member for around ten years and has taught several core courses in pure and applied mathematics at undergraduate, postgraduate, and doctorate levels. He is currently a Professor at Graphic Era (Deemed to be University), Dehradun, India. Before joining the Graphic Era, he was a Deputy Manager (Probationary Officer) with Syndicate Bank for a short period. He is Editor-in-Chief of *International Journal of Mathematical, Engineering and Management Sciences* and the Guest Editor & Member of the editorial board of various journals. He is a regular Reviewer for international journals, including IEEE, Elsevier, Springer, Emerald, John Wiley, Taylor & Francis and many other publishers. He has published 125 research publications in IEEE, Taylor & Francis, Springer, Elsevier, Emerald, World Scientific and many other national and international journals of repute and also presented his works at national and international conferences. His fields of research are reliability theory and applied mathematics. Dr. Ram is a Senior Member of the IEEE, life member of Operational Research Society of India, Society for Reliability Engineering, Quality and Operations Management in India, Indian Society of Industrial and Applied Mathematics, member of International Association of Engineers in Hong Kong, and Emerald Literati Network in the U.K. He has been a member of the organizing committee of a number of international and national conferences, seminars, and workshops. He has been conferred with "Young Scientist Award" by the Uttarakhand State Council for Science and Technology, Dehradun, in 2009. He has been awarded the "Best Faculty Award" in 2011 and recently Research Excellence Award in 2015 for his significant contribution in academics and research at Graphic Era.

Contributors

Rajeev Agarwal

Department of Management Studies
Malaviya National Institute of Technology
Jaipur, India

Kumar Anupam

Department of Chemical Engineering
Deenbandhu Chhotu Ram University of
    Science and Technology
Haryana, India

Shwetank Avikal

School of Engineering and Technology
Graphic Era Hill University
Dehradun, India

Gaurav Kumar Badhotiya

Department of Mechanical Engineering
Graphic Era (Deemed to be University)
Dehradun, India

Wen-Kuo Chen

Department of Marketing and Logistics
    Management
Chaoyang University of Technology
Taichung, Taiwan

Pankaj Kumar Goley

Chemical Recovery and Biorefinery
    Division
Central Pulp and Paper Research Institute
Uttar Pradesh, India

Tanuja Gour

School of Management
IMS Unison University
Dehradun, India

Ashish Gupta

Department of Marketing
Indian Institute of Foreign Trade
    (IIFT)
New Delhi, India

Naveen Jain

Department of Mechanical Engineering
Shri Shankracharya Institute of
    Professional Management and
    Technology
Raipur, India

Sunil Kumar Jauhar

Operations Management and Decision
    Sciences
Indian Institute of Management Kashipur
Kashipur, India

Kishori Kasat

Symbiosis Institute of Computer
    Studies and Research
Symbiosis International (Deemed
    University)
Pune, India

Vimal Kumar

Department of Information Management
Chaoyang University of Technology
Taichung, Taiwan

Kuei-Kuei Lai

Department of Information Management
Chaoyang University of Technology
Taichung, Taiwan

Balkrishna Eknath Narkhede

Industrial Engineering and
    Manufacturing Systems Group
National Institute of Industrial
    Engineering (NITIE)
Mumbai, India

Vaibhav S. Narwane

Department of Mechanical Engineering
K. J. Somaiya College of Engineering
Mumbai, India

Millie Pant
Applied Science and Engineering
Indian Institute of Technology
Roorkee, India

Rakesh D. Raut
Operations Management Group
National Institute of Industrial
    Engineering (NITIE)
Mumbai, India

Naim Shaikh
Global Business School and Research
    Centre
Dr. D. Y. Patil Vidyapeeth
Pune, India

Monica Sharma
Department of Management Studies
Malaviya National Institute of Technology
Jaipur, India

Nagendra Kumar Sharma
Department of Business Administration
Chaoyang University of Technology
Taichung, Taiwan

Vinay Sharma
Department of Production Engineering
Birla Institute of Technology Mesra
Ranchi, India

Mahesh Shinde
Symbiosis International
    (Deemed University)
Pune, India

A. R. Singh
Department of Mechanical Engineering
National Institute of Technology Raipur
Raipur, India

Amar Singh
School of Commerce
Graphic Era Hill University
Dehradun, India

Rohit Singh
Systems Engineer
Infosys Limited
Bhubaneswar, India

Pragati Sinha
Department of Management Studies
Malaviya National Institute of Technology
Jaipur, India

Mayuresh S. Suroshe
K. J. Somaiya College of Engineering
Mumbai, India

Prashant Tiwari
Institute of Business Management
GLA University
Mathura, India

Shiv Kant Tiwari
Institute of Business Management
GLA University
Mathura, India

Neha Verma
Department of Mechanical Engineering
Shri Shankaracharya Institute of
    Professional Management and
    Technology
Raipur, India

Pratima Verma
Department of Information Management
Chaoyang University of Technology
Taichung, Taiwan

Anil Yadav
Engineering and Maintenance Division
Central Pulp and Paper Research
    Institute
Uttar Pradesh, India

1 Sustainability in the New Normal

Industry 4.0 Perspective

Naim Shaikh[1], Mahesh Shinde[2], and Kishori Kasat[3]

[1]Global Business School and Research Centre, Dr. D. Y. Patil Vidyapeeth, Pune, India
[2]Symbiosis International (Deemed University), Pune, India
[3]Symbiosis Institute of Computer Studies and Research, Symbiosis International (Deemed University), Pune, India

1.1 INTRODUCTION

Industrial development has significantly benefited the various stakeholders in-vovled. However, evidence shows that the manufacturing methods adopted by most industries are not environmentally and ecologically sustainable (McWilliams et al., 2014). Therefore, there is a constant demand for industrial production to be eco-nomically, socially and environmentally sustainable (Sarkis and Zhu, 2017).

Industrial sustainability calls for adoption of the principles of allocating natural resources rationally, waste reduction, and environmental and ecological conserva-tion. At the same time, there is a wave of consumer awareness based on sustain-ability concepts resulting in a sustainable consumption pattern (Terlau, Wiltrud, and Hirsch, 2015).

Energy use, carbon emissions and resource utilization are the three main areas that the manufacturing sectors around the globe are focusing on for achieving in-dustrial sustainability. Industrial sustainability is no longer a buzzword, rather a strategy to gain a competitive advantage. This advantage can be drawn by optimally using resources, including energy, and thus directly influencing the triple bottom line.

Sustainable industrial development will call for an innovative business strategy, completely altering the current business models and relationships and offering novel products and services. To achieve this "Industry 4.0" presents an innovative business model based on cyber physical systems, an integration of human effort and technology primarily focusing on reducing wastage and achieving the utmost po-tential of industrial processes. Industry 4.0 can be perceived differently by different industries or even within a industry. This meaning is dependent upon the strategy undertaken by that industry. However, the common elements in all this are technological connectivity of manufacturing reducing time and intelligence gap and most importantly operational excellence.

Connectivity plays an important role in adoption and implementation of industry 4.0. This connectivity has a positive impact on the entire supply chain. This will result in on demand-customised production, thereby reducing the inventory cost and wastage related to overstocking of inventory: a step ahead of just in time. Industry 4.0 will allow autonomous decision-making, enabling optimum utilization of resources, high quality at affordable cost reducing wastage, and thus catering to sustainable industrial development. This will be achieved by decentralised co-ordination of all the stakeholders across the value chain and synchronization of all industrial processes at one-time decision-making precision.

Switching over from manually controlled operation management to automation powered by Industry 4.0 in the manufacturing sector will allow efficient resource utilisation, flexibility in operation, quality consciousness—and at a reduced cost. Industry 4.0 thus provides a solution in the VUCA world for aligning the business agenda with the Agenda 21 of sustainable development. This also affords an opportunity to reduce the lead-time of production and achieve cost saving in the mass production process. This is easier said than done and requires primary focus on the right priorities and then integrating appropriate Industry 4.0 solutions to achieve the sustainable development goals. The solutions definitely would be unique to a particular industry and will require a specific industry-oriented approach (Figure 1.1).

Goal 9 of Sustainable Development Goals speaks about industry innovation and infrastructure. It also points out importance of industrial infrastructure to address all sustainable development issues. It also provides for importance of technology and innovation to achieve in closed energy and resource efficiency. Development is directly related to industrialisation, which ultimately depends upon technology and innovation. Industry 4.0 provides a wonderful opportunity of achieving all these goals. This chapter analyses how Industry 4.0 solutions can be utilised for achieving these goals.

1.2 DEFINING SUSTAINABILITY

Defining industrial sustainability is not an easy task. The definition changes with reference to time, place, and creates a major problem in aligning uniform sustainable solutions. Both environmental and ecological sustainability and ecological problems are dynamic and very much reactive to the state of affairs. These concepts are construed differently by different people in different parts of the world and even differently by same people in different perspectives and circumstances. Information dissymmetry is one of the key reasons for this disparity and differences of opinion. This is further compounded by sphere of vision of the stakeholder, technological awareness and costs associated.

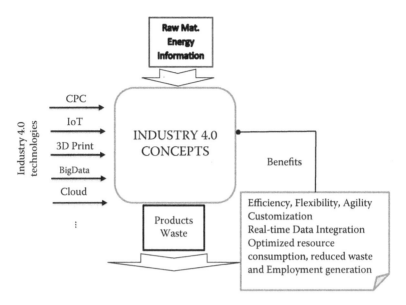

FIGURE 1.1   Benefits of industry 4.0.

The second contributory factor is the value position of the stakeholders, particularly in terms of knowledge about ecosystems and their commitment towards sustainable goals. Diverging value positions and opinions are direct consequences of incomplete information about environment and different levels of commitment towards sustainability. This creates a problem of generating a uniform solution rather defining the sustainability in a uniform commonly applicable way. Consequence being development of interesting views and opinions about sustainability and related to those diverse definitions of ecological and environmental problems and sustainable solutions for the same.

This manifests into different ways in which people define environmental problems and similar situation being interpreted in a different way depending on the location of the stakeholder and the period referred. This is the precise reason why there are different definitions of sustainability, rendering all attempts to create a universally applicable definition of sustainability a futile exercise.

One of the accepted and dominant views regarding sustainability is that of defining sustainability in terms of planetary boundaries. This concept revolves around the belief that the planet on which we are living and are dependent on has its own limits and these limits can be quantified based on the urgency. This approach is popular because it accepts there is a lack of available information, regarding boundaries, at the current moment and these cannot be clearly defined. This idea has allowed the construction of a concept of boundaries from particular to general increasing the limits of boundaries with reference to the perspectives of the stakeholders. Thus, at a primary level, there could be an individual boundary wherein the individual actions and the consequences of the same on the environment could be considered. At a slightly higher level, it would be a social boundary consisting of interaction between people at societal level and consequences on the environment. At a slightly higher level, it could affirm an industry's relationship with its stakeholders affecting the sustainable environment in different ways and paving the path for creating sustainable products and processes.

Because of this, the popular boundaries drawn by people revolve around the following aspects: Regional cluster, which talks about an industry or a group of industries and their impact on environment. At a slightly higher level, comes the bottom set of city boundary where in rather than considering the impact of an industrial production, a complete system view of production and consumption pattern of the city is taken into consideration. This can be further extended to the whole earth system and enlarged to the concept of universe.

So, while applying sustainable solutions, all agents should be at the same boundary level or the solutions offered and the perspectives considered will not match with each other. Alignment of system boundaries to create uniform sustainable solutions is a primary requirement for success of this approach and defining sustainability in this perspective. Even at this level, there exists huge information dissymmetry and thus prohibits creation of constructive ideas and definition. In a nutshell the definition of sustainability is person-specific depending upon, "Who is analysing and what boundaries the analysis is restricted to?" Alignment of system thus plays a critical role in definition of sustainability. Brundtland commission report provided encompassing definition of the system for sustainability:

"Sustainable development is development that meets the needs of the present without compromising the ability of future generations to meet their own needs. So this is about development, it's about development in the present which also takes into a ground what is possible in the future." *"Our Common Future: Report of The World Commission On Environment And Development*, 1987.

This definition is not restricted to the rights of the present generation only but also considers the rights of future generations on the natural resources. The main source for evolution of the concept of sustainable development has been in the practice of forestry. Forestry teaches about harvesting capacity, sustainable yield and sustainable utilisation of forest resources.

Sustainability redefines the basis of man-nature relationships. It goes further and talks about equality and equal access to resources not only to the present generations but also to the future generations. Finally, the concept sustainable development does focus on three aspects and nature of industrial activity on these three pillars: economic, ecological and social.

1.3 RELATIONSHIP BETWEEN SUSTAINABILITY AND TECHNOLOGY

Managing industrial production within the ecological constraints, ensuring the use of natural resources at par with the rate of regeneration, reducing waste generation, and ensuring biosphere conservation are few of the mandates of environmental sustainability (Herman, 1999). The idea of industrial development automatically draws attention to the concept of the ecological impact of industrialization. Technology then is regarded as unavoidable when considered as one of the major causes of ecological impact on environment. However, a different perspective shows that technology can also provide solutions to the ecological impact and seen from this perspective, technology upgradation should be the endeavour to reduce the ecological impact of there is a technology and improve the efficiency misconception that development associated with technology is antithesis to sustainability. This general trend of surrendering to the technology and accepting what technology offers needs to be replaced by an element of choice allowing to choose the type of development that technology is offering. This presents society with an opportunity of alternatives and choosing the best technology that will offer development with improved environmental performance and sustainability. The cascading effect of this would result in pursuit for search of sustainable technologies and associated sustainable development, altering the common belief of the polar relationship between technologies empowered development and sustainability.

In conclusion the changed perspective would view technology not as something inevitable and detrimental to the environment but as a wilful choice for sustainable development. This requires a serious change in management position regarding managing responsibility. Sustainability in the industrial sphere then revolves around the three pillars of responsible management, ethics and, most importantly, technological solutions.

Implementation of responsible management practices would require adoption of sustainable management tools at systems level. These tools can be classified into two types: Outcome Oriented Tools and Process Oriented Tools. Outcome oriented tools indicate the level environmental impact of activities or level of achievement of sustainability and include ecological foot printing, ecosystem services, life cycle assessment, environmental impact assessment among the others. On the other hand, process-oriented tools rather than calculating the amount of environmental impact focus on the way in which this impact could be assessed. Examples of processoriented tools include Environment Management System, which is also covered by ISO certification.

There are plenty such tools available. The choice of these tools will depend upon the particular industry, timeframe and locational setting of the industry. This confusion is further supplanted by the local legal systems, dilemma between sustainable cost and profits, amount of financial power, will of the management, technological awareness, customer awareness and vigil of the society to name the few. This creates dilemma regarding choice of the tool at an industrial level and begins the search for all encompassing solution.

1.4 INDUSTRY 4.0

Industry 4.0 provides an umbrella solution for sustainability practices, which would be accepted at all levels and boundaries. Industry 4.0, based on collective and cooperative endeavour between man and technology and powered by technologies like cyber physical systems, internet of things, industrial internet of things, big data analytics, virtual reality, augmented reality, robotics and other advanced technologies offer tremendous scope on significantly contributing to sustainability along with environmental protection. The fourth industrial revolution has transformed the industrial production sector leading to digitalization, intelligent decision making and technological reconnected decentralized manufacturing (Semel, Wagemann and Herrmann, 1996)

Industry 4.0 can significantly contribute to the sustainable development goals impacting industrial production systems and processes like conservation of raw material and resources, reduction of waste and costs associated with waste disposal, energy conservation, increasing efficiencies by automated and decentralized machine based decision-making processes, which will increase the overall efficiency while reducing costs. Positive adoption of Industry 4.0 based philosophy will allow transformation of industrial business processes, making them a resource that is more efficient, conservative and cost effective. This, in turn, will help the industrialists, employees, stakeholders, society and government to achieve sustainable development goals.

The philosophy of industry 4.0 revolves around the theme of operational flexibility in manufacturing efficiency and sustainability, ensuring constantly high quality with low cost along with profits offering customised eco-friendly products and services to the customers (Wang et al., 2015).

Technological advancement, cyber physical systems with IOT representing smart factories, products, and integration of vertical and horizontal business processes across the value chain are based on technologies like autonomous robots, industrial IOT cyber security, simulation, cloud additive manufacturing, big data, augmented and virtual reality. This allows increase autonomy and flexibility coupled with virtual industrial manufacturing practices and real-time decision making, which can support sustainable manufacturing goals (Industry 4.0: The Future Of Productivity And Growth In Manufacturing Industries M Rüßmann, M Lorenz, P Gerbert, M Waldner, J Justus... - Boston Consulting Group, 2015).

Industry 4.0 is is a philosophy enabling sustainable development (Beier, Niehoff and Xue, 2018). Ample literature is available showing the link between industry 4.0 technologiesand industrial sustainability. Industry 4.0-allied business processes ensure the sustainability of power use. (Ghobakhloo, 2019).

Industry 4.0 also provides business models, which are primarily based on sustainability and resource conservation (Cornelis de Man and Strandhagen, 2017).

Industry 4.0 is also and enabler of lean manufacturing practices (Kamble, Gunasekaran, and Dhone, 2019).

Sustainability and Industry 4.0 have various common considerations across the value chain. These may include practices like design for manufacturing life cycle management lean and green management safe working conditions eliminating waste and improving efficiencies (Machado, Winroth and Ribeiro da Silva, 2019).

Industry 4.0 has significant potential for contributing towards sustainable manufacturing and development goals by creating smart business models, additive manufacturing powered product and process design that is sustainable, IT enabled decentralized quicktime decision making providing resource efficiency, creating positive impact on product life cycle and in general creating synergies in all cradle-to-cradle activities (Stock and Seliger, 2016).

Significant contribution of industry 4.0 in manufacturing as cited in literature is as under:

1. Highly customised output DANIEL KIEL et al., "SUSTAINABLE INDUSTRIAL VALUE CREATION: BENEFITS AND CHALLENGES OF INDUSTRY 4.0", *International Journal of Innovation Management* 21, no. 08 (2017): 1740015, doi: 10.1142/s1363919617400151.
2. Reduced inventory and increase agility resulting in positive impact on lead-time creating value across the chain. Bressanelli, Gianmarco, Federico Adrodegari, Marco Perona, and Nicola Saccani. "Exploring how usagefocused business models enable circular economy through digital technologies." *Sustainability* 10, no. 3 (2018): 639.
3. Life cycle assessment analysis creation of efficient business models based on innovation and value creation.
4. Positive integration of data creating environment of learning and development.
5. Allowing simulation and technology enabled processes.
6. Optimal integration of vertical and horizontal processes and digital integration of business processes. *Challenges and Solutions for The Digital Transformation and Use of Exponential Technologies; Deloitte: London,* 2015.

1.5 ANALYSIS

Undoubtedly, Industry 4.0 has a significant potential in promoting sustainability but there are other impacts that need to be evaluated before coming to conclusion. Following are possible impacts that Industry 4.0 can have on sustainability at various stages of manufacturing.

1.5.1 ADOPTION AND FORMATION STAGE

Cyber physical system powered Industry 4.0 technologies will require procurement of new plant and machinery, integration of the existing machinery with cyber physical system using sensors and high value software solutions. As opposed to the concept of conservation of resources, this digital transformation will indirectly lead to enhanced consumption of natural and human resources. Condemnation and disposal of old obsolete plant and machinery equipment and systems hardware and software air will increase resource wastage (Müller, Buliga and Voigt, 2018).

This digital transformation will require catering ad-ditional budget for pro-curement of robotic and electronic devices and sensors, which will further increase extraction of rare minerals like lithium and heavy earth materials thus depleting the natural resources *(Marscheider-Weidemann et al., 2016; German Mineral Resources Agency, 2016).*

Process automation will have its consequences like enhanced demand for new plant and machinery and increased quantity of obsolete plant and machinery thus negatively affecting the sustainable development goals by increased demand for new plant and machinery and necessary energy demands for transportation and costs towards disposal. Similarly, digital transformation will also have same impact where in there will be increased demand for rare earth minerals, increased energy consumption levels and fuel demands for transportation, increased levels of obsolesce, and associated disposal costs. Digital integration will also increase the demand for new equipment and enhanced condemnation of obsolete old plant and machinery having similar necessary negative consequences on sustainability.

The above-mentioned problems provide challenges for adoption and implementation of Industry 4.0 as a sustainability provider industrial tool. This calls for all stakeholders to focus on innovation and provide thought on creating alternate strategy, which will reduce natural resource consumption and substitute wastages.

1.5.2   OPERATION STAGE

Industry 4.0 empowered digitalisation backed by IOT and big data analytics will allow decentralized and independent decision-making with the aid of technology reducing human error and allowing predictive maintenance, better control of production processes and overall management of factory production system. However, this is is not without a negative impact. Cyber physical systems, industrial internet. of things and real time decision-making and data processing will demand a tremendous amount of data, which will lead to increase energy consumption (Stock et al., 2018).

However, there is another side to this. Industry 4.0 and associated digitalization of production processes will create opportunities for power consumption, improved energy efficiency, and at the same time, optimal use of resources. Decentralized decision making powered by technology we love error-free quick decisions bring in reduction of waist and improved efficiencies. Technologies like robotics and additive manufacturing flexible manufacturing creating customised on-the-market products with energy-efficient production processes in the long run (Sreenivasan, Goel and Bourell, 2010).

Cyber physical system-based smart production facilities powered by IOT and real-time decision making have several positive impacts on sustainability, including horizontal and vertical integration of business processes and customer centric customised production processes. These allow optimal use of raw material and energy, recycling of resources and providing reliable data about consumption patterns of consumers as well as energy to manage life cycle issues of production. On demand and customised production further eliminates waste, decrease raw material and power consumption and extend life cycle of products.

Big data analytics optimizes consumption of raw material and energy allows predictive and preventive maintenance and thus reduces the overall cost. Technologies like 3D printing decrease wastage by allowing prototyping simulation and reducing the overall costs. However, at the same, time they need higher power consumption.

1.5.3 INDUSTRY 4.0 AND SUSTAINABILITY

Industry 4.0 definitely provides for various sustainable solutions. However, adoption of this technology is a cost-intensive process and this cost involves financial as well as environmental repercussions. It thus becomes essential to properly plan and coordinate the transformation of the current production systems to Industry 4.0 (Bonilla et al., 2013).

If integrated in a judicious manner, Industry 4.0 has the potential to contribute to several sustainable development goals. Industry 4.0 and associated combination of cyber physical systems with industrial internet of things and data analytics allow decreased energy consumption, increased renewable energy use, decreased waste generation, increased and enhanced life cycle of products, and decreased greenhouse gas emission thus allowing affordable and clean energy, industrial innovation and infrastructure, responsible consumption and production and reducing adverse impact on global climate change.

The cost-benefit analysis of Industry 4.0 vis-a-vis sustainability has to be done keeping in mind the long-term perspective. Undoubtedly, in a long run it appears that it requires initial cost for adoption and implementation of this philosophy. However, at present, the philosophy is in a nascent stage and requires elaboration and expert expansion. Long-term risk benefit analysis provides that even with the initial cost associated in the long run, Industry 4.0 has a tremendous potential to create a competitive advantage for businesses, ensuring sustainable value creation (Müller, Kiel and Voigt, 2018).

If due diligence is observed in adoption and implementation of this philosophy, it will have a significant impact on production systems by reducing resource demand and raw material consumption, by allowing recycling and substitution, satisfying customers who are now demanding environment-friendly sustainable products, provide job potential and upliftment of skills of workforce and will create value and work as a facilitator for achieving sustainable development goals (Ford and Despeisse, 2016).

1.5.4 INDUSTRY 4.0 AND ECONOMIC SUSTAINABILITY

Cyber physical system powered Industry 4.0 allows cost saving, efficient use of energy and resources by creating infrastructure and equipment at lower cost through appropriate use of technologies like industrial internet of things, artificial intelligence, machine learning, big data analytics, 3D printing, augmented reality and virtual reality. Use of a proper combination of these technologies will allow continual endeavour to reduce operational cost. Vertical and horizontal integration of business processes will convert the supply chain into value chain. This will aid in industrial planning, reduced maintenance cost, and increased efficiency.

1.5.5 INDUSTRY 4.0 AND SOCIAL SUSTAINABILITY

At social level, industry 4.0 will bridge the gap between technology and human effort. It will get synergized industrial output through combined efforts. It will allow a low-cost high-quality manufacturing, which is sustainable and energy efficient. It will allow upgradation of the skill sets of labour, thus improving the scale of operation of the labour. It will also create a culture of 360 degree learning and lifelong education. It will allow creation of green and lean processes with lesser carbon emissions satisfying diverse stakeholders.

1.5.6   INDUSTRY 4.0 AND ECOLOGICAL SUSTAINABILITY

On the ecological front, Industry 4.0 through its technology-based interventions helps in manufacturing, reduction of industrial wastes, and improves the optimal use of natural resources. In the analysis of life cycle of the product Industry 4.0 interventions based on using sensors help in mapping emission levels, key performance indicators, behavioural aspects as well as improvements in process models. This allows decentralized decision making and robotic intervention in analysis, prevention and correction in the product life cycle issues as well as offers transparency, which allows a competitive advantage to the business as well as eco-friendly sustainable products to the customers.

1.6   CONCLUSION AND DISCUSSION

The major concern of all the stakeholders globally including the governments, non-government organisations, customers, media and industries is determining modalities for fulfilment and satisfaction of sustainable development goals. The dimensions of these needs range from efficiency in production, reduced consumption of natural resources, sustainable use of natural resources, alternate renewable use of material and energy, energy conservation, poverty alleviation to social sector applications like creating jobs unemployment. Technology-driven Industry 4.0 is looked upon by almost all stakeholders as an umbrella solution which satisfy all these needs.

Businesses are optimistic about the long-term benefits of Industry 4.0 across the three pillars of sustainability: economic social and ecological. Economically, Industry 4.0 is expected to reduce the pressure on the raw material used, reduce the lead times and create sustainable production systems based on efficient use of energy. On the social side, Industry 4.0 is expected to create a healthy and safe working environment, a 360-degree learning culture, and upgradation of skills of the workforce. On the ecological front, the major issues are use of raw material, waste generation and efficient use of non-renewable sources, particularly energy.

These benefits are expected of creating strategic competitive advantage for the businesses in terms of costs savings on raw material and energy conservation, legal compliance and improved processes which lead to improved quality and customer satisfaction. These benefits are not without the flip side of the perspective. Adoption and implementation of Industry 4.0 has various inherent concerns which need to be addressed.

On the economic front, Industry 4.0 is a cost intensive venture. It also requires high level of technical expertise at all levels of employee hierarchy and needs to be supported with technological co-ordination amongst all the verticals and across the entire value chain. Cyber physical systems and other pillars of Industry 4.0 based on the latest technology do present serious cost considerations to begin with. On environmental platform adoption of industry 4.0 will lead to purchase of new technology complaint plant and machinery including investment on software technologies which will further put pressure on use of natural resources. This will also lead to obsolesce of existing plant and machinery t leading to issue of waste disposal and also disposal of electronic waste that would be generated due to intensive use of computer technology. Socially the entire concept of Industry 4.0 is based on the interaction coordination and synchronisation between the effect of technology and humans it inherently poses threat to the employees with low skill levels and also creates fears about loss of jobs. Another aspect affecting the social acceptance of this technology is the data security and privacy issues.

In conclusion it would be appropriate to say that the Industry 4.0 technologies at inception stage involve cost, technology adaptation and change in organisational culture affecting almost the entire hierarchy. It will also affect the sustainability negatively to a certain extent because of investment required for this switch over. It will also put forward pressure on natural resources of different nature required for this technology creating issues of e waste disposal and disposal of obsolete plant and machinery. On a social side Industry 4.0 is also looked upon as employment disrupter.

But at the same time in the long run, Industry 4.0 has immense potential of overcoming all these short-term hiccups. Used in a diligent manner, Industry 4.0 promises on all the fronts to achieve sustainable development goals and create the future production facilities that are economically and ecologically sustainable and socially acceptable.

The purpose of this chapter was to present a bird's eye view of the impact Industry 4.0 can have on sustainable development goals. The chapter did not intend to go through deeper analysis of the Industry 4.0 technologies and the consequent impact on sustainability.

REFERENCES

Beier, Grischa, Silke Niehoff, and Bing Xue. (2018) "More Sustainability in Industry Through Industrial Internet of Things?" *Applied Sciences* 8(2), p. 219. doi: 10.3390/app8020219.

Bonilla, Silvia H., Cecília M.V.B. Almeida, Biagio F. Giannetti, and Donald Huisingh. (2013) "Key Elements, Stages and Tools for a Sustainable World: An Introduction to This Special Volume". *Journal of Cleaner Production*, 46, pp. 1–7. doi: 10.1016/j.jclepro.2012.12.011.

Bressanelli, Gianmarco, Federico Adrodegari, Marco Perona, and Nicola Saccani. (2018) "Exploring How Usage-Focused Business Models Enable Circular Economy Through Digital Technologies." *Sustainability*, 10(3), p. 639.

Brundtland Commission. (1987) Our Common Future: Report of the World Commission on Environment and Development.

Daly, Herman E. (1999) "From Uneconomic Growth to a Steady-State Economy". Cheltenham, U.K.: Edward Elgar Publishing Ltd.

Deloitte. (2015) "Challenges and Solutions for the digital transformation and use of Exponential Technologies", Deloitte, London, U.K.

Ford, Simon, and Mélanie Despeisse. (2016) "Additive Manufacturing and Sustainability: An Exploratory Study of the Advantages and Challenges". *Journal of Cleaner Production*, 137, pp. 1573–1587. doi: 10.1016/j.jclepro.2016.04.150.

Ghobakhloo, Morteza. (2019) "Determinants of Information and Digital Technology Implementation for Smart Manufacturing". *International Journal of Production Research*, 58(8), pp. 2384–2405. doi: 10.1080/00207543.2019.1630775.

Kamble, Sachin, Angappa Gunasekaran, and Neelkanth C. Dhone. (2019) "Industry 4.0 and Lean Manufacturing Practices for Sustainable Organisational Performance in Indian Manufacturing Companies". *International Journal of Production Research*, 58(5), pp. 1319–1337. doi: 10.1080/00207543.2019.1630772.

Kiel, Daniel, Julian M. Müller, Christian Arnold, and Kai-Ingo Voigt. (2017) "Sustainable Industrial Value Creation: Benefits and Challenges of Industry 4.0". *International Journal of Innovation Management*, 21(8), 1740015. doi: 10.1142/s1363919617400151.

Machad, Carla, Gonçalve Mats, Peter Winroth, and Elias Hans Dener Ribeiro da Silva. (2019) "Sustainable Manufacturing in Industry 4.0: An Emerging Research Agenda". *International Journal of Production Research*, 58(5), pp. 1462–1484. doi: 10.1080/00207543.2019.1652777.

Marscheider-Weidemann, Frank, Sabine Langkau, Torsten Hummen, Lorenz Erdmann, and Luis Tercero Espinoza. (2016) "Raw Materials for Emerging Technologies". Berlin, Germany: German Mineral Resources Agency.

McWilliams, Abagail, Annaleena Parhankangas, Jason Coupet, Eric Welch, and Darold T. Barnum. (2014) "Strategic Decision Making for the Triple Bottom Line". *Business Strategy and The Environment*, 25(3), pp. 193–204. doi: 10.1002/bse.1867.

Müller, Julian Marius, Daniel Kiel, and Kai-Ingo Voigt. (2018) "What Drives the Implementation of Industry 4.0? The Role of Opportunities and Challenges in the Context of Sustainability". *Sustainability*, 10(1), 247. doi: 10.3390/su10010247.

Müller, Julian Marius, Oana Buliga, and Kai-Ingo Voigt. (2018) "Fortune Favors the Prepared: How SMEs Approach Business Model Innovations in Industry 4.0". *Technological Forecasting and Social Change*, 132, pp. 2–17. doi: 10.1016/j.techfore.2017.12.019.

Rüßmann, Michael, Markus Lorenz, Philipp Gerbert, Manuela Waldner, Pascal Engel, Michael Harnisch, and Jan Justus. (2015) "Industry 4.0: The Future of Productivity and Growth Industries, Boston Consulting Group, Boston, MA, USA 2015.

Sarkis, Joseph, and Qingyun Zhu. (2017) "Environmental Sustainability and Production: Taking the Road Less Travelled". *International Journal of Production Research*, 56(1-2), pp. 743–759. doi: 10.1080/00207543.2017.1365182.

Semel, J., K. Wagemann, and K. Herrmann. (1996) "48. SUSTECH - eine Initiative der Europäischen chemischen Industrie zur Forschungskooperation". *Chemie Ingenieur Technik*, 68(9), pp. 1094–1095. doi: 10.1002/cite.330680950.

Sreenivasan, R., A. Goel, and D.L. Bourell. (2010) "Sustainability Issues in Laser-Based Additive Manufacturing". *Physics Procedia*, 5, pp. 81–90. doi: 10.1016/j.phpro.2010.08.124.

Stock, T., and G. Seliger. (2016) "Opportunities of Sustainable Manufacturing in Industry 4.0". *Procedia CIRP*, 40, pp. 536–541. doi: 10.1016/j.procir.2016.01.129.

Stock, Tim, Michael Obenaus, Sascha Kunz, and Holger Kohl. (2018) "Industry 4.0 as Enabler for a Sustainable Development: A Qualitative Assessment of Its Ecological and Social Potential". *Process Safety and Environmental Protection*, 118, pp. 254–267. doi: 10.1016/j.psep.2018.06.026.

Terlau, Wiltrud, and Darya Hirsch. (2015) "Sustainable Consumption and the Attitude Behaviour-Gap Phenomenon-Causes and Measurements Towards a Sustainable Development." *International Journal on Food System Dynamics*, 6(3), pp. 159–174.

Wang, Shiyong, Jiafu Wan, Di Li, and Chunhua Zhang. (2016) "Implementing Smart Factory of Industrie 4.0: An Outlook". *International Journal of Distributed Sensor Networks*, 12(1), 3159805. doi: 10.1155/2016/3159805.

2 Sustainability of Large-Scale Industries in the Global Market

Pragati Sinha, Monica Sharma, and
Rajeev Agrawal
Department of Management Studies, MNIT Jaipur, India

2.1 INTRODUCTION

2.1.1 SUSTAINABLE DEVELOPMENT – *THE ORIGIN*

The deterioration in the environmental and human conditions globally, the growing challenges in reducing poverty in low-income countries, the incessant struggle to support developing economies without ignoring environmental threats, and the failure to address lopsided development of industrialization called for the formation of the Brundtland Commission, formally known as the World Commission on Environment and Development (WCED) in 1984 (Arena et al., 2009). The Brundtland Commission, an organization independent of the UN, was established to unite countries and build a consensus to focus on identifying similar issues, raising awareness about holistic growth, and suggesting solutions that aid the implementation of sustainable goals. In 1987, the Brundtland Commission submitted its first volume of the report titled "Our common future", which gave the now most prevalent definition of sustainable development:

Sustainable development is development that meets the needs of the present without compromising the ability of future generations to meet their own needs. It contains within it two key concepts:

- the concept of "needs", in particular the essential needs of the world's poor, to which overriding priority should be given; and
- the idea of limitations imposed by the state of technology and social organization on the environment's ability to meet present and future needs".

The Brundtland Commission has been successful in cultivating shared values between governments and organizations, with the biggest success reflected by the wide acceptance of sustainable issues by global players since 1987 (Holden et al.,

DOI: 10.1201/9781003102304-2

2014). The acceptance of the novel concepts from the report discussed during the first Rio Earth Summit in 1992 marked the beginning of a new era where scientists, policy experts, government leaders, and the bourgeoisie took part in thinking, re-thinking, and strategizing a new pathway for world leaders (Barrutia et al., 2015). This report simplified the understanding of sustainable progress by providing a three-pillar framework. The three main pillars of sustainable development include economic growth, environmental protection, and social equality. Among these, environmental protection has been widely investigated.

2.1.2 Industrial Sustainability – An Antidote to Environmental Degradation

The purpose of economic development in any country is to provide improved living and jobs. Although large-scale industries have contributed to economic develop-ment by generating employment and removing poverty, enabling a higher output of consumer goods at cheaper rates, and raising the standards of living (Industrial Development Report, 2018), the negative results of such industries cannot be ig-nored. The rapid disappearance of natural resources; the rising levels of pollution (air, water, noise, and soil); the continuously increasing energy consumption; the increasing cases of diseases due to dust, smoke, and harmful particles; waning cottage industries; and declining opportunities for regional and local artisans have resulted in a lopsided economic development (Stock & Seliger, 2016). A pollution-free industrial development is a myth – it is neither possible nor required. What is required is adopting a balanced approach with a greater focus on least-polluting production methods, higher utilization of industrial waste, and reduction of income disparities (Zaccai, 2012). The onus (social and environmental) rests on the highest-polluting industries, such as thermal power plants, coal mines, cement, sponge iron, steel & ferro alloys, petroleum and chemicals, fashion, and retail.

The acknowledgement of sustainable industries has led to the development of industrial manufacturing processes in a sustainable manner (Arena et al., 2009). Industrial Sustainability (IS) refers to the end state of a transformation process where industry acts as both a component of and an active contributor to a socially, environmentally, and economically viable and sustainable planet (Tonelli et al., 2013). So, how does Industrial Sustainability help remove the pain points associated with economic development? Large-scale industries are advised to adopt the 3P framework by addressing three pain points – environmental pollution, social in-equalities, and slow economic growth (Szolnoki et al., 2014) (Figure 2.1).

The pictorial depiction of the framework highlights the integration of eco-nomically viable, environmentally benign, and socially beneficial parameters that aid the multicriteria decision-making procedure of selecting sustainable options for manufacturing. However, it becomes imperative to address the barriers and challenges associated with achieving a perfect model of sustainable large-scale manufacturing. Commitment to strengthen the circular economy (CE) and revisit industrial manufacturing strategies provides a solution to the problems associated with the "take-make-waste" extractive industrial model (Xavier et al., 2019).

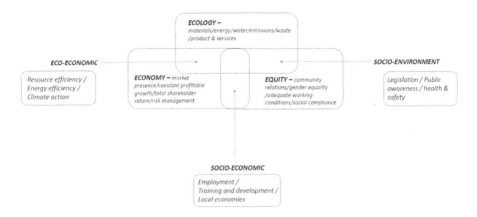

FIGURE 2.1 3E model.

Large- and medium-scale industries need to invest in firming up the circular economy model. The EU proposed the concept of CE in 2007 which was accepted by several governments, including China, Japan, The UK, France, Canada, The Netherlands, Sweden and Finland, as well as by several businesses around the world. China was the first country to adopt CE as a law in 2008 (Sutherland et al., 2020). A circular economy model typically focuses on an economic system of closed loops in which raw materials, components, and products lose their value as little as possible and where renewable energy sources are used to their full potential. Despite the known benefits of the circular economy and CE being the most viable alternative for the linear economy, the first Circulatory Gap Report, on CE published in 2018 showed disheartening signs of low acceptance in a real-world working scenario. The report states that the world economy is only 9.1% circular, leaving a huge circulatory gap. This hints at a potential market segment that remains untapped. Industries need to adopt a systematic method to design for durability, re-utilization, remanufacturing, and recycling. There remains a scope to boost the four R concept – Reduce, Reuse, Recycle, and Refurbish. Additionally, large-scale organizations need to adopt B2B and B2C marketing methods that promote responsible consumption and raise consumer awareness about the benefits of the circular economy.

2.1.2.1 Relevance of Industrial Sustainability Across Industries

The fashion industry is one of the most polluting industries across the world. The textile industry is the second-largest pollutant of local freshwater and contributes to one-fifth of the water consumption among all industries (Colucci et al., 2020). The Environmental EPA Cotton contributes to 16% of pesticide pollution. The life cycle of fashion products, cradle to death, is marked by air, water, soil pollution and huge piles of discarded clothes that end up in landfills. The take-make-discard model propagated by fast fashion has received lot of criticism from environmentalists and forced big fashion brands like Zara and H&M, to develop strategies to support the 3P business model.

The mining industry has been known to be dominated by waste generation and exploitation of natural resources (Xavier et al., 2019). Many experts have actively voiced their concern about the negative impacts of the mining industry due to the continuous extraction of non-renewable resources. Mining activities produce large amounts of mining waste during the beneficiation, extraction, and processing of minerals. Extraction results in wastes such as debris, soil, and other components that are unusable in the mining industry (Abramov et al., 2020). While land mining has long been a concern due to the associated environmental and health degradation, along with waste generation, deep-sea bed mining, for the purpose of extracting rare metals, has recently gathered environmentalists' (Le et al., 2017). The mined sea bed is more vulnerable to climate change. Carbon emission, air pollution (by toxins such as mercury, lead, Sulphur, nitrogen dioxide), contamination of groundwater, depletion of natural resources, deforestation, soil erosion, pollution of water bodies, impact on aquatic animals, and loss of biodiversity are some of the direct and indirect impacts of mining on the environment at the local, regional, and global levels. Besides occupational hazards, the health impact of the mining industry is alarming; most miners develop respiratory or skin diseases at some point in their life (Xavier et al., 2019).

The chemical industry is another significant source of environmental pollution. Chemical manufacturing processes have a threefold impact on the environment and human health. Chemical industrial processes result in harmful emissions into the air and water, generating solid and slurry waste that pollutes the soil. Another major concern is the continuous extraction of natural resources at every stage of manufacturing, which leads to the exhaustion of natural resources. Moreover, the chemical industry indirectly charges the environment with emissions of sulfur dioxide, nitrogen oxides, and particulate matter (Burton et al., 2019).

Although electricity is a safe and clean form of energy consumption, the process of electricity generation in any form leads to air, water, and land pollution (Mascarenhas et al., 2019). The major challenge with power generation is the production of carbon dioxide (CO_2) – which results in greenhouse gas; nitrogen oxide (NOx) – which contributes to ground-level ozone generation, which, in turn, can cause lung irritation; and sulfur dioxide (SO_2) – the primary reason for acid rain (Bilgen, 2014).

The indirect role played by industries like consultancy, retail, and logistics can also be harmful in terms of carbon footprint and this has been studied in detail by environmentalists (Tonelli et al., 2013).

2.1.3 INDUSTRY 4.0 AND CIRCULAR ECONOMY – *ALIGNING TECHNOLOGY AND SUSTAINABILITY*

Smart factory, smart products, smart business models, and smart consumers – there is a remarkable notional change in the way consumer expectation and product offering has initiated the alteration of market fabrication. The emerging fourth industrial revolution, often referred to as Industry 4.0, involves fast and disruptive changes due to digital manufacturing, network communication, computer, and automation technologies.

This revolution is often referred to as Industry 4.0, because it aims to not only compound the relevance of smart operations but also to strongly impact the whole value chains and provide a set of new opportunities, in terms of business models, production technology, creation of new jobs, and work organization (Pereira & Romero, 2017). This new industrial paradigm embraces a set of technological developments, such as CPS, IoT, Robotics, Big Data, Cloud Manufacturing, and Augmented Reality, that enhance the efficiency and productivity of both products and processes (Parida et al., 2019).

Innovation and technology are the answer to develop organizational and operational systems of sustainable production and consumption, focused on restoring the value of used resources (Sandstrom, 2020). Countries have realized the importance of extending the life cycle of products, creating opportunities for the reuse of raw materials, smart operational methods to remanufacture and recycle products (Sutherland et al., 2020). Industry 4.0 supports a paradigm shift from the linear to the circular economy by focusing on smart and efficient production methods by computerization of manufacturing (Ng & Ghobakhloo, 2020). Though the relationship between CE and Industry 4.0 has not been explored widely, there are pertinent questions that have been repeatedly brought up for analysis:

1. What technologies and resources from Industry 4.0 need to be identified to best suit the advancement of CE?
2. A thorough planning process with an intense study of demand and supply to both cater and encourage responsible consumption habits.
3. Establish a roadmap to shorten the lead time in establishing CE principles in organizations?
4. How can Industry 4.0 be adopted to achieve sustainable operation management?
5. How to strengthen SSCM – Sustainable Supply Chain Management?
6. Identify barriers to smooth functioning.

2.1.4 Sustainable Development Goals (SDG) – *Inclusive and Sustainable Development by Industries (ISID-I)*

In September 2015, 193 member States of the United Nations met in New York to adopt 17 new Sustainable Development Goals ("SDGs"). The common mission of SDG 2015–2030 is to meet the 3P framework and make the world more prosperous, inclusive, sustainable, and resilient (Alcorta Ludo, 2017). Businesses cannot achieve SDG by themselves. The role of the government in helping businesses meet sustainable goals is supposed to be both directive and supportive. The success achieved by industries in fulfilling SDG has been supported by policies and procedures formulated by governments. SDG goals 7, 8, 9, 12 and 13, in particular, are the greatest opportunities for industries to contribute towards sustainable growth (refer to the SDG goal chart). The below table highlights some of the milestones achieved by large-scale industries in meeting the SDG goals (Table 2.1).

TABLE 2.1
SDG Goals and Industry Contribution

SDG Ref#	SDG – description		Contribution by Industry
1	End poverty everywhere	Ecology	Create manufacturing sink in untapped regions
		Equity	Creating jobs and opportunities for all / food distribution / equitable growth
		Economy	Increase the proportion of products that are sourced and manufactured locally in developing and emerging economies / Invest in R&D to design and produce for developing nations

Practiced by: ACC (India) – treats local communities as stakeholders and nurtures them / DBL group (Bangladesh) – established fair price shop for low waged workers / Maruti Suzuki (India) – localized sourcing, 78% of sourcing from local suppliers in the vicinity of manufacturing plant

SDG Ref#	SDG – description		Contribution by Industry
2	Zero hunger	Ecology	Increasing energy efficiency/ providing machinery and tools to agriculture that reduces farming hazards and optimizing water utilization/collecting and analyzing data from sensors and adjusting heating, ventilation, light and irrigation accordingly
		Equity	share knowhow with farmers in extended supply chain and treat farmers as valuable resource
		Economy	procure biomass from farmers and aid additional source of earning

Practiced by: Ambuja cement (India) – purchases biomass from farmers / Jain irrigation systems (India) – designed drip irrigation system for smallholder farmers / Royal Philips – partnered with universities to develop indoor farms using LED lights / ThyssenKrupp Industries India, with its large market presence in the sugar industry, is constantly looking to develop more energy-efficient equipment and processes for all areas of sugar production.

SDG Ref#	SDG – description		Contribution by Industry
3	Good health and well-being	Ecology	Strict measures to curb harmful effects of emissions (in air and water) that cause diseases and disorders
		Equity	Improve working conditions / provide healthcare insurance and facilities, feeding room and clean toilets for women / physical and psychosocial support after disaster events
		Economy	Manufacture and sale of medicines and medical devices tailored to the needs of middle and low-income countries

Practiced by: General Electric – healthcare business of GE has incubated and spun off a start-up company in India for the first time, attempting to create an ambitious healthcare distribution network

TABLE 2.1 (Continued)
SDG Goals and Industry Contribution

SDG Ref#	SDG – description	Contribution by Industry	

for Tier-2 and Tier-3 towns and cities as part of a push to solve some of the biggest healthcare challenges facing the country / Hitachi – Hitachi Proton Beam Therapy System is one example of healthcare innovation to apply advanced technology in accelerators, irradiation, and control systems.

SDG Ref#	SDG – description	Contribution by Industry	
4	Quality education	Ecology	R&D and innovation to encourage sustainable designing and production of eco-friendly commodities
		Equity	Skill based training programs for lower segments to enable equal job opportunities / collaborate with other businesses, NGOs and governments to improve learning in countries within the company's value chain
		Economy	Provide training to component and raw material suppliers to increase the productivity and sustainability of their operations, ensuring access to high quality, environmentally sensitive inputs

Practiced by: General Electric – healthcare business of GE has incubated and spun off a start-up company in India for the first time, attempting to create an ambitious healthcare distribution network for Tier-2 and Tier-3 towns and cities as part of a push to solve some of the biggest healthcare challenges facing the country / Hitachi – Hitachi Proton Beam Therapy System is one example of healthcare innovation to apply advanced technology in accelerators, irradiation, and control systems.

SDG Ref#	SDG – description	Contribution by Industry	
5	Gender equality	Ecology	Evidence shows that advancements in gender equality could have a profoundly positive impact on social and environmental well-being – women are active agents of conversation and restoration
		Equity	Adapt manufacturing plant facilities, processes and culture to support an increase in recruitment, development and retention of women employees / Increase the share of women on company boards and in senior roles
		Economy	Encourage women participation in businesses /

Practiced by: DBL Group implemented the "Women in Factories Initiative" program of Walmart with technical support from the NGO CARE to create awareness about rights of women at workplace/ Tata Steel has created a "Women Empowerment Cell", comprising members of management and junior staff, including unionized female employees for women from disadvantaged backgrounds.

(Continued)

TABLE 2.1 (Continued)
SDG Goals and Industry Contribution

SDG Ref#	SDG – description		Contribution by Industry
6	**Clean water and sanitation**	Ecology	Reduce water consumption – closed loop manufacturing processes and replacing wet-machining with dry-machining processes / Improve water treatment facilities, reduce emission of harmful substance in water / invest in water recycling programs / Sign the WASH pledge of the World Business Council for Sustainable Development which calls on companies to implement access to safe water, sanitation and hygiene at the workplace.
		Equity	Integrate community access to water within production facility design / Develop and manufacture reliable, low-cost water pumps and sanitation technology which are adapted to the needs of low-income communities in rural areas and to high density urban areas
		Economy	National economies are more resilient to rainfall variability, and economic growth is boosted when water storage capacity is improved

Practiced by: Ambuja Cements Ltd has been working on Water Resource Management since 1993 to enhance water resources in communities around all its plants across India (Initiatives include its Roof Rain Water Harvesting System, micro-irrigation, interlinking of canals and mined out pits) / Daimler AG has introduced a "zero discharge" policy in its new plant in Chennai in southern India. It channels water through a complex system of pipes, pumps, filters, and evaporators in a closed loop and it is continually reconditioned, with no water leaving the plant via a sewer line / Hitachi manufacturers water treatment equipment for use in plants, factors and power stations, to ensure wastewater from industrial production does not cause pollution, and it is treated and recycled in a closed system to conserve diverted water.

SDG Ref#	SDG – description		Contribution by Industry
7	**Affordable and clean energy**	Ecology	Develop more efficient microgrid technology capable of integrating renewable energy sources to bring affordable, renewable energy to rural and marginalized communities / Develop and market industrial machinery and vehicles, vessels and aircraft that run efficiently on sustainable energy sources
		Equity	Encourage and support suppliers to increase the proportion of their energy coming from renewable sources / Promote local suppliers of solar panels
		Economy	Develop energy infrastructure and technologies that make renewable energy (e.g., solar and

TABLE 2.1 (Continued)
SDG Goals and Industry Contribution

SDG Ref#	SDG – description		Contribution by Industry

wind) a more compelling economic proposition by increasing reliability, increasing storage capacity and reducing cost) / Renewable energy improves human well-being and overall welfare well beyond GDP. Doubling the share of renewables in the global energy mix increases global GDP in 2030 by up to 1.1%, equivalent to USD 1.3 trillion.

Practiced by: Airbus Group is supporting the development of sustainable fuels made from biomass feedstock that, through their lifecycle, emit less CO2 than conventional fossil fuels / • Hitachi has applied its IT expertise to contribute to the development of infrastructure which provides a stable energy supply. In many countries, grid operators are pursuing more robust wide-area interconnection with the aim of liberalizing energy markets and improving reliability. High-voltage direct current (HVDC) transmission systems convert electricity to direct current before transmission, reducing electricity losses, facility sizes, and construction costs, while expanding access to electricity nationwide

SDG Ref#	SDG – description		Contribution by Industry
8	Decent work and economic growth	Ecology	Integrate small-scale producers of component parts into the supply chain that take care of sustainable parameters
		Equity	Local sourcing and manufacturing in low and middle-income countries, where viable, to reduce extreme poverty and lift the local economies / Equal health opportunities for all – Promote high standards of health and safety in manufacturing facilities and extraction sites, Invest in technologies that reduce the risk of human error and accidents in production / Create opportunities for lower-paid workers to develop their skills and gain access to improved professional opportunities / Set supplier standards that require suppliers to uphold labour rights (including equal opportunities, equal pay for equal work, rights of migrant workers, and safe working conditions) and support their implementation through supplier training and monitoring
		Economy	Provide targeted internships for young people from disadvantaged backgrounds in order to promote social mobility whilst also enhancing

(Continued)

TABLE 2.1 (Continued)
SDG Goals and Industry Contribution

SDG Ref#	*SDG – description*	*Contribution by Industry*
		company performance through increased workforce diversity /

Practiced by: Ford has implemented a training program to promote responsible working conditions in its supply chain / Hewlett-Packard (HP) expanded its supply chain requirements in 2014, taking major steps toward preventing exploitative labour practices and forced labour / Hyundai supports small-scale suppliers in its supply chain as part of its pursuit of mutual growth, Hyundai also extends loans and other financial support to suppliers, In addition, Hyundai extends voluntary technical guidance and support developed for Tier 1 suppliers to smaller Tier 2 suppliers / Vedanta Resources focuses on hiring, developing and retaining talent from local communities.

SDG Ref#	*SDG – description*		*Contribution by Industry*
9	**Industry innovation and infrastructure**	**Ecology**	Invest in R&D to create more environmentally sensitive construction material and develop improved methods of reusing by-products and waste /
		Equity	Innovate products to meet the needs of rural areas / collaborate to provide infrastructure for rural development
		Economy	Create industrial zones with Govt to encourage complementary investments in infrastructure, technology and production / Develop innovative financing strategies (affordable microloans) / Engage with governments in high-growth markets to discuss ways in which more sustainable building products, transportation solutions and manufacturing techniques can help develop local infrastructure and economies

Practiced by: Atlas Copco is one of the largest producers of air compressors and industrial facilities can typically save up to 10% on their total energy bills by using its compressors / MAN SE, a German mechanical engineering company, has launched its first carbon-neutral commercial vehicle assembly plant, which runs exclusively on regenerative energy / Tata Steel has been aggressively working to develop LD slag as a legitimate, cost effective and, most importantly, green product

SDG Ref#	*SDG – description*		*Contribution by Industry*
10	**Reduced Inequalities**	**Ecology**	Create job centres in rural areas – Encourage handicraft and regional craft that develop ecolabels
		Equity	Adopt equal opportunity policies prohibiting discrimination in all forms / encourage women entrepreneurs / Create opportunities for lower-paid workers to develop their skills and gain access to improved employment opportunities

TABLE 2.1 (Continued)
SDG Goals and Industry Contribution

SDG Ref#	SDG – description		Contribution by Industry
		Economy	Pay staff a living wage and encourage other companies within the value chain to also pay living wages

Practiced by: Volkswagen AG is committed to supporting employees with performance impairment or disabilities / Siemens AG has adapted its German apprenticeship program to accommodate young people with below-average school performance or who lack basic competencies, often due to their background as migrants / Cemex provides all its employees around the world with a living wage, whilst protecting their labor rights in a safe and respectful work environment

SDG Ref#	SDG – description		Contribution by Industry
11	Sustainable cities and communities	Ecology	Develop products that improve the energy efficiency of people's homes and offices includes lighting, ventilation, heating and air-conditioning / Develop innovative, low-cost construction materials / Develop and market more sustainable transport solutions, particularly public buses and trains / promote natural or green infrastructure to create societal and ecosystem value
		Equity	Industrial dispersal – create labour requirement and service need / collaborate with small businesses, they have a big impact on the economy, both on a national and local level
		Economy	Collaborate with government and other companies reduce employment – invest in greater connectivity, energy efficiency and safety to strengthen urban communities

Practiced by: Broad Group has developed a new, cheaper form of steel structure which enables rapid construction of high-rise buildings whilst also improving interior air quality / Cemex's flagship inclusive business, Patrimonial Hoy, was founded to provide low-income families with access to affordable housing by providing finance, building materials, technical advice and logistical support, enabling them to build or expand their homes more quickly and efficiently.

SDG Ref#	SDG – description		Contribution by Industry
12	Responsible consumption and production	Ecology	Factor an internal carbon price into capital project decisions / reduce fossil fuel combustion in industrial manufacturing plants / Increase energy efficiency in industrial manufacturing plants and across distribution networks / Source materials with lower embedded energy
		Equity	Promote homegrown ecofriendly brands / collaborate with craft sector that supports women livelihood (fashion industry) /

(Continued)

TABLE 2.1 (Continued)
SDG Goals and Industry Contribution

SDG Ref#	SDG – description		Contribution by Industry
			collaborate with social institutions to provide job opportunities for disabled
		Economy	Apply 4R (reduce, reuse, recycle, refurbish) concept to promote circular economy

Practiced by: Apollo Tyres is investing in innovative technologies to increase the sustainability of its tyres / Fuji Xerox Co., Ltd. operates a "closed loop" integrated recycling system for its products, in which products released to the market are collected back after use, and the parts are either reused or recycled, thus reducing waste sent to landfill / Hyundai targets an 85% recycling rate for the plastic, rubber and glass in its end-of-life vehicles, and a 95% recovery rate.

SDG Ref#	SDG – description		Contribution by Industry
13	Climate action	Ecology	Design and implement natural disaster risk mitigation / Set science-based carbon emission targets in line with the sectoral decarbonization pathway and encourage suppliers, distributors and customers to do the same
		Equity	Support climate justice – 70% of population in developing countries reside in regional areas vulnerable to climatic hazard. Climate justice ensures safeguarding the less privileged
		Economy	Strengthen economy by supporting high level partnerships and industry associations advocating for responsible public policies on modernization with focus on climate

Practiced by: Hyundai undertakes diverse activities to develop ecofriendly cars and reduce the amount of greenhouse gases created during the manufacturing of vehicles / Caterpillar is innovating to improve and build products that are both valuable to its customers and more sustainable.

SDG Ref#	SDG – description		Contribution by Industry
14	Life below water	Ecology	Design pumps and other machinery for deep sea mining which minimize the risk of marine spillages and contamination / Design components for marine vessels that minimize the risk of marine pollution / Collaborate with other stakeholders to collect and utilize marine plastic waste / Implement improved waste treatment systems / Ensure supplier and distributor companies shipping goods by sea adhere to environmental standards on marine shipping
		Equity	Ensure diversity and gender inclusiveness at all levels, to set a balanced course for humanity and foster innovative solutions for the ocean (UN at worlds ocean day emphasized that women are engaged in all aspects of

TABLE 2.1 (Continued)
SDG Goals and Industry Contribution

SDG Ref#	SDG – description		Contribution by Industry
			interaction with our ocean, yet their voices are often missing at the decision-making level)
		Economy	Marine ecosystems around the world provide a wealth of ecosystem services (the benefits people obtain from nature), including food provision for billions of people, carbon storage, waste detoxification, and cultural benefits including recreational opportunities

Practiced by: Interface, a carpet tile manufacturer, has a carpet tile collection called Net Effect that honors the ocean through project Net-Works – Net-Works provides a source of income for small fishing villages in the Philippines, while cleaning up their beaches and waters.

SDG Ref#	SDG – description		Contribution by Industry
15	Life on land	Ecology	Use wood from certified sustainable sources / Manufacture printers with environmental features / Develop and manufacture soil-friendly technology
		Equity	Farmland preservation helps to ensure sufficient farmland to supply communities with locally grown produce and helps farmers improve their economic well-being
		Economy	Industrial support in preserving life on land has socioeconomic impacts – land is one of three major factors of production in classical economics (along with labor and capital)

Practiced by: Caterpillar joined an effort in 2015 to focus on restoring natural infrastructure – the forests, prairies, farmlands, wetlands and coastal landscapes / Mitsui & Co., Ltd. owns forests at 74 locations throughout Japan / Tata Steel has applied innovative bioengineering methods to reduce runoff from its mining facilities / Xerox's 2020 goals include helping to preserve the world's forests and biodiversity.

SDG Ref#	SDG – description		Contribution by Industry
16	Peace, justice and strong institution	Ecology	Identify and assess risks of conflict minerals in supply chains by identifying suppliers of 3TG metals (Tin, Tantalum, Tungsten and Gold) and designing a process of necessary due diligence for those suppliers
		Equity	Design and implement a robust anti-bribery and corruption compliance program / Demonstrate ethical leadership by publishing a statement on human rights consistent with the UN

(Continued)

TABLE 2.1 (Continued)
SDG Goals and Industry Contribution

SDG Ref#	*SDG – description*		*Contribution by Industry*
			GuidingPrinciples on Business and Human Rights and sign up to the ten principles of the UN Global Compact
		Economy	Improving traceability of products, parts and materials in the supply chain to ensure reliability of sustainability claims covering human rights, labor, anti-corruption and the environment

Practiced by: Fluor, a global engineering construction company, supports external anti-corruption efforts through collective action (Fluor works on zero tolerance towards bribery of any form even if it will lose business or encounter delays because of its refusal to do so) / Hyundai promotes fair trade practices

SDG Ref#	*SDG – description*		*Contribution by Industry*
17	**Partnerships for the goals**	Ecology	Engage in multi-stakeholder initiatives – develop supply chain of environment pro suppliers and vendors
		Equity	Strengthen the link between corporate and societal value creation / Adopt good practice principles and guidelines which better align business practices with sustainable development and development for all segments
		Economy	Establish a robust impact measurement framework for corporate, multi-stakeholder partnership and industry level contributions to sustainable development including regular monitoring and transparent evaluation and reporting.

Practiced by: Companies within the industrial manufacturing industry have collaborated with each other and with additional stakeholders to develop several good practice initiatives and collaborations.

2.1.5 SUSTAINABLE GLOBAL MARKET (SCOPE AND POTENTIAL) – ROLE OF INDUSTRY 4.0

The Business & Sustainable Development Commission launched its *Better Business, Better World* report, at Davos in 2016. The sustainable business market was valued at at least $12 trillion a year (by 2030). In fact, total sustainability investments across sectors and industry types actually exceed some of the fastest-growing sustainability markets. Today, total sustainable investments top up to $30 trillion – up 68% since 2014 and tenfold since 2004. Another market sentiment assessment indicator is the Green Technology and Sustainability Market. (The chain process of structures and systems, products and services, application tools, and course of action that aim to conserve the environment- and nature-bestowed resources are collectively called Green Technology, for example, solar cells). As per reports, the green technology and sustainability market size was valued at $6.85 billion in 2018, and is projected to reach $44.61 billion by 2026, growing at a CAGR of 26.5% from 2019 to 2026. The global green technology and sustainability market is studied by segmenting the market as per region, technology, and application. The industry is predicted to generate a revenue of $57.8 billion in 2030 against $8.3 billion in 2019. Additionally, between 2020 and 2030 (forecast period), the market would advance at a CAGR of 20.0%. (Allied market research report, 2020) (Figure 2.2).

Many industries have started adopting green technology as technological transformation and diversification methodology for meeting the sustainable development goals and repositioning them as responsible manufacturers (Allied market research report, 2020). The Internet of Things (IOT) is valued as the most promising segment followed by cloud computing and AI. An increase in research and development (R&D) activities and growing demand for RFID sensors that eliminate carbon emissions and support clean energy is observed as the current trend in large-scale industries (Stock & Seliger, 2016). Large manufacturing units have realised that it is time to stop perceiving sustainability as cost-centric, and instead, treating "green manufacturing" as a core value to generate revenues and earn loyal

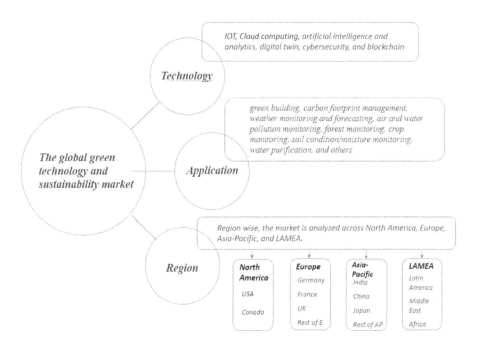

FIGURE 2.2   Segmentation of global green and technology market.

customers. Sustainability is a tool to compound revenue and profitability along with securing competitive advantages (Bandyopadhyay et al., 2017). CropX Inc., Enablon France SA, Enviance Inc., General Electric, Hortau Inc., IBM Corporation, LO3 Energy, Inc., Oracle Corporation, Tech Mahindra Limited, and Trace Genomics, Inc are a few prominent contributors to the green technology and sustainability market (Allied market research report, 2020). The innovations in technology to aid manufacturing, the intense competition in the dynamic market and the multilayered strategies to win consumers are going to magnify with increasing requirements for business survival.

2.1.5.1 Digital Technologies Need to Create a Resource Efficient Industrial Base

It is expected that by 2050, the manufacturing output will increase four times more than what it was in 2017 (Mckinsey report, 2020). Organizations need to ensure that this quadruple increase in goods and services is achieved, while aiming for zero waste, zero climate change, lower levels of emissions, and utilization of half of the current resources. The significance of using digital technologies in developing a resource efficient industrial base is the need of the hour. Industries adopt digital technologies to (Parida et al., 2019):

1. Identify the base-level changes in a firm to enable performances measurable by sustainability indices.
2. Adapt processes to machine components and robots that optimize the consumption of materials and energy.
3. Direct investment in R&D to develop manufacturing processes to considerably reduce emissions and improve energy efficiency.
4. To ascertain definitively the demand patterns and existing opportunities for sustainable processes in industries.
5. To develop business models that bring a shift towards a circular economy.

2.1.5.2 Benefits of Technologically Advanced Sustainable Industrial Manufacturing Processes

Cyber-physical systems, cloud manufacturing, IOT, and additive manufacturing enable smart operating systems due to a high level of interconnectedness of machines, orders, employees, suppliers, and customers. Decentralized and automated manufacturing systems allow components, digital devices and machines to communicate with each other and enable self-managed production lines (Demartini et al., 2019). The entire supply chain is smart – from design to production, distribution and retail. The integrated, adapted, optimized, and interoperable manufacturing processes not only minimize errors but majorly reduce wastages. The availability of real-time data that enables real-time monitoring, controlling, and recording of important parameters in traditional operational procedures can now be applied to monitor sustainable parameters. Production parameters such as energy consumption, efficient allocation of natural resources, flow of materials, customers' expectation for traceability, and maintaining suppliers' data can be monitored and controlled with the input of cyber-physical manufacturing system (CPMS) (Pereira

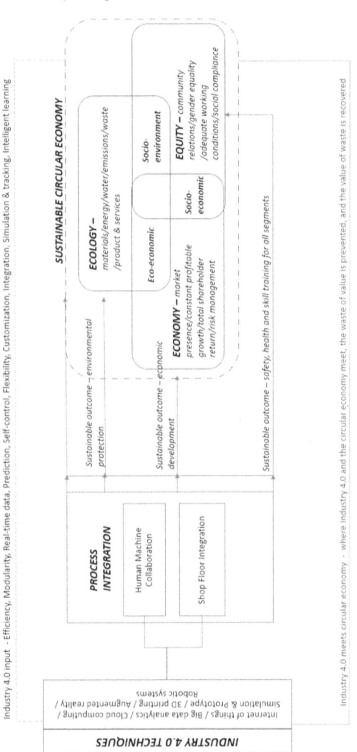

FIGURE 2.3 Industry 4.0 techniques applied to sustainability model.

& Romero, 2017). Let us understand the application of industry 4.0 in the context of sustainable manufacturing (Figure 2.3).

Industry 4.0 includes connectivity, advanced analytics, automation, and advanced manufacturing techniques. This has been accepted worldwide, and it has started helping companies achieve product efficiency, customization to meet consumer expectation, reduce time lag in production, service effectiveness, and error minimization. These techniques, to a large extent, lead to a circular and sustainable business model (Abramov et al., 2020). The Industry 4.0 model framework is essentially determined by horizontal integration across the value chain network, end-to-end engineering across entire product life cycles, and vertical integration and networked manufacturing systems. The network of value creation factors: equipment, human, materials, organization, and processes lead to new business models backed by data precision that ensure longer product timelines, zero defects, and zero wastage (Tonelli et al., 2013). The four foundational technologies applied along the value chain are:

1. Connectivity, data, and computational power
2. Analytics and intelligence
3. Human-machine interaction
4. Advanced engineering

Data infrastructure, analysis, and intelligence (AI/ML) and process automation impact the entire supply chain: demand sensing, efficiency in material procurement, optimization of resources during manufacturing, reduction in factory-to-shelf timeline and sales planning. The optimization of resources during manufacturing results in lesser extraction and burden on natural resources; demands sensing results in accuracy in demand planning that results in shrinking inventory; robotics and automation reduce error and wastage – Industry 4.0 techniques pave way for a sustainable business model (Tonelli et al., 2013). Additionally, Industry 4.0 also aids a circular business model (CBM), where Industry 4.0 meets CBM, the value of waste is recovered: it addresses the end phase life cycle of product containing reuse, remanufacturing, recycling, recovery and disposal.

2.1.6 Challenges in Achieving Industrial Sustainability

The holistic goal of Industrial sustainability is to align the industrial processes with the responsibility of minimum extraction of natural resources (energy, water, materials), developing a closed loop supply chain that is value driven from design to consumption and to concentrate on generating equal opportunities in society. Investing in sustainability never guarantees immediate returns and since it is a long and less attempted journey, highlighting the challenges and identifying the barriers becomes imperative (Fröhling & Hiete, 2020).

2.1.6.1 A Large Segment of Unaware and Unprepared Consumers

Consumers seem to be ready to embrace sustainability. However, most research shows that they actually do not end up buying sustainable products regularly. This "intention-action" gap has to be narrowed down for any corporate sustainable program to be successful (Industrial Development Report, 2018). Eminent players of the industry need to come together to create a strategy to spread awareness among consumers about the benefits of sustainable living. The onus of nurturing a sustainable supply chain (SSCM) rests equally on the end-user. Unilever estimates that 70% of the greenhouse footprint depends on which products customers choose and whether or not they follow sustainable methods to dispose them (HBR report, 2019). Acknowledging and emphasising the role of marketers in moulding consumer behaviour in favour of green consumption is an important integration while strategizing the green supply chain. Harnessing the power of social media in cultivating "green thinking as a culture" must be on top of most marketing strategies. Sample studies in multiple markets have shown that once consumers are made aware of best practices and eco-friendly products, organizations are pressured into embracing sustainable practices.

2.1.6.2   The Stakeholders

In simple words, stakeholders are parties that have an interest in the company and either affect or get affected due to the business. Investors, employees, suppliers, wholesalers, distributors, retailers and consumers are stakeholders that influence the strength of a sustainable operating system (Arena et al., 2009). The two main challenges faced by industries in adopting sustainable manufacturing processes is the underdeveloped eco-system that struggles to support an efficient and consistent closed-loop green supply chain and the non-rewarding government policies in most countries that are yet to favour Industrial Sustainability. The industry has been forced to evolve in the past in response to new regulations, technologies, and changing customer demands – every stakeholder needs to co-create demand and supply of products and services that lead to ecological and social balance. The governments of developed and developing nations need to reform the decision-making processes to allow more integrative approaches that aid transparency and public participation at all levels.

2.1.6.3   Competitiveness and Lock-in Period

Few dedicated players and low competitiveness in a relatively slow-growing sustainable market pose a challenge to establish sustainable competitiveness. Competition augments efficiency and fuels innovation to develop a wider and deeper product range. Sluggish competition or few powerful players both dominate and limit the market and prevent better quality, better-priced products, and services. Competition policy is yet to play a significant role in achieving sustainable and inclusive growth and development. To add on, the longer lock-in period for returns in sustainable manufacturing further prevents an individual, a firm, or an industry to sustain or increase the ability to generate income in relation to the current and future wider environment and society (Alcorta Ludo, 2017).

2.1.6.4   Organizational and Product Fit

Industry 4.0 cannot be implemented in isolation; it requires synchronization and coordination with existing equipment and processes, which involves complexities and a high level of investment. With doubts regarding the organizational preparedness for implementation of industry 4.0 techniques and related new technologies, integrating sustainable parameters adds to the existing challenges. Organizations are continuously under stress from governments, consumers, employees, and investors to demonstrate that they are adopting ethical and sustainable business practices (Spoelstra, 2013). The cost-benefit analysis makes it difficult for organizations to choose ethical business practices and give up short- term profit maximization goals that pull continuous investment. Moreover, there is no "one-size-fits-all" approach to creating a sustainable approach to business in different contexts. Examples with regard to the challenge of organizational and production fit are automotive, electrical engineering, and steel industry sectors, which struggle to implement Industry 4.0, let alone achieving Industrial sustainability.

2.1.6.5 Limited Potential of Recycling and Substitution – Choosing the Right Materials Throughout the Product Life Cycle

"Product lifecycle sustainability" is an approach to managing the stages of a product's existence so that any negative impact on the environment is minimized. Continuous monitoring of environmental footprint of materials used and products manufactured from beginning till end of product life cycle, in tandem with return, reutilization and remanufacture of products demands accuracy and never-ending standardization and control mechanism. At each stage of industrial manufacturing, natural resources are consumed and emissions to air, water and soil are released in the environment. Life-cycle assessments at each stage (design-extraction of raw materials-manufacturing-packaging-distribution-end consumption-recycling) in large scale industries are complicated and require lot of detailing (Sutherland et al., 2020).

2.1.6.6 Employee Skill Set – How Prepared Are Employees to Imbibe Sustainability as Part of Tech-Operated Culture?

The implementation of Industry 4.0 concept requires the top management to educate and connect the employees with the process of value creation (Parida et al., 2019). Management activities and processes that do not breed futuristic mindset fail to tap the real potential of human resource. Managers should not only be eager to convince employees to address the benefits of industry 4.0 but actively develop specific competencies and skills, in data analytics, IT, software, and human-machine interaction but also treat employees as the first set of social partners in creating a green supply chain. Both workers and management must work towards energy saving processes and reduce environmental impact starting from workplace. Technical training for environmental trainers who work as green representatives; constant strengthening of employee capacity to tackle climate change issues at firm and industry level, workplace-based training initiatives, conferences and workshops, sectoral social dialogue to ensure a level of skills in new green technologies among employees are few initiatives that have not been aggressively taken by organizations, except in Europe (Sandstrom, 2020). Virtual reality, augmented reality training to fresh recruits on shop floor is the new age training requirement.

2.1.7   ROLE OF GOVT – FOCUS ON NATIONAL SDG, INCH CLOSER TOWARDS GLOBAL SDG

The industries cannot achieve the Sustainable development goals (SDG) by themselves, the role of government is pertinent in developing policies, review "green business" framework and develop dialogues with large and medium scale organizations to control the procedures that impact business and environment (Lin et al., 2011). Government must motivate manufacturers to look beyond philanthropic short-term goals and CSR activities, rather the manufacturers should be guided to pin the SDG goals 7, 8, 9, 12 and 13 as a window for biggest contributions. According to PwC 19th Annual Global CEO Survey, 69% of CEOs confirm that businesses will be adhering to SDGs and also that encouragement from Governments shall have a higher influence in fulfilling SDG agendas by 2030. However, 79% of CEOs expressed concern that business growth may be hampered due to Government regulations and interference, and hence the need to adopt a fine balance. No single govt can develop policies and frameworks in favor of socio-economic-ecological balance and develop a macroenvironment in favor of green supply chain when the govt worldwide work in opposite direction (Tonelli et al., 2013). The government needs to factor the global guidelines in domestic policy agenda. Industries (large or medium scaled) technologically advanced and determined to support sustainable will fail in a society that is not ready for sustainable consumption – efforts to create a movement will fall short in absence of an unaware and unprepared audience. Governments need to set economic agendas that are based on fair and transparent practices. In absence of strong government support, companies fail to withstand the competition and loose opportunities to scale the sustainable business model.

Despite the commendable start in 1987 after Brutland report and the echo of common sentiments at the Earth Summit at Rio, the post summit era saw governments struggle with assuming responsibilities in the new role (Holden et al., 2014). By 1990s government role was dominated by deregulation, downsizing, and deficit reduction.

With the advent of recent volatile markets, it becomes imperative for governments to encompass collective decisions made in public sector, private sector and civil society. The merger of public-private sectors to co-create policy instruments (economic, social, and informative) is essential as an alternative to regulations and restrictions that limit the scope of exploring the sustainable markets. Some of the short- and long-term government roles highlighted as below:

- collaboration with stakeholders (national and international),
- partnership with other governments for common agenda of sustainability,
- implementing local SDG,
- creating best mix policies, with focus to encourage innovation
- green fiscal audits (green budgets and ecological fiscal reform),
- steering society through nation-wide awareness campaigns towards sustainability,
- sustainability practiced in government operations and purchasing policies,

- encouraging manufacturers for R&D to produce eco-friendly goods
- government-wide information for awareness, training and technical assistance to enable execution, advancement of indicators to monitor and evaluate programmes and policies (including reverse supply chain)

2.1.8   INDUSTRIAL SUSTAINABILITY – COVID-19 AND WAY FORWARD

The global production and supply chain network are affected due to Covid-19. Both big and small plants are grappling to manage operations to meet dwindling demand pattern. Even in parts of the world where Covid has started to recede, the large-scale industries are struggling to cope with serious dislocations, worker unavailability and increasing pressure of reduced demand (Kumar et al., 2020). The most voiced out question to be answered right now is: How will manufacturing and its supply chains look after COVID-19? Businesses need to be much more digital and highly connected for quick decision making as a response to changed scenario due to the pandemic. According to the report "Industry 4.0: Reimagining manufacturing operations after COVID-19" by Mckinsey published in June 2020, Industry 4.0 solutions have been implemented by Industry leaders. As per the Mckinsey report, 39% have executed a nerve center, or control tower to enhance transparency in supply chain. The reports also suggest that automation has been adopted for fast tracking programs to cope with worker shortages due to COVID-19. While world leaders in manufacturing have started restructuring the operating systems, it is time to intensely integrate potential of industry 4.0 to resolve the problem of waste and inefficiency. It is imperative for manufacturing industry to take cognizance of the fact that sustainable practices or circular economy is not mere about resource conservation, but mainly about how linear turned circular economies can recover value from waste, along with preventing loss of resources. This fourth industrial revolution supports circular business models that only consume renewable resources and keep materials from infinite stocks to infinite loop (Xavier et al., 2019).

Operational Technology

There is an urgent need for an agile and flexible approach to develop business model that focuses on closed-loop green supply chain with increased planning for scrap recycling (Lin et al., 2011). Industries need to focus on Standardization of environmental performance – especially maximizing scrap value and deploying technological breakthrough in upcycling industrial scrap. Also, while most large-scale industries focused on globalized supply chain pre COVID-19, there is a magnified requirement to geo-relocate in coming years. The key learning from this worst-ever pandemic is to keep the production supply chain flexible and resilient with an increasing focus on nationalized and localized sourcing and manufacturing methods and reduced dependency on outsourcing. The growing sense of nationalism has to be propagated by suppliers, manufacturers and all stakeholders. Collaboration and partnerships to deploy technology to upgrade operational systems at local manufacturing centres, investments in homegrown start-up that provide solutions for circular supply chain and intensified digital solutions for demand sensing are a few major strategic implementations post-pandemic.

Information Technology

data collection is at the core of all business practices, products that are connected to the IOT allow manufacturers to control and analyze their performance from a distance and collect continuous data for further process improvements (Grappi et al., 2017). This lays the foundation for many circular business models (CBM) and a must have for all companies that seek answer to post COVID-19 performance enhancement. With growing need to control fluctuations in demand and provide opportunity to produce durable products to control waste, manufacturing units require flawless equipment and machineries. This can only be done when manufacturers are able to observe and investigate performance at a distance. There is also a dire need for manufacturing units to adopt Recycling and Upcycling, remanufacturing and re-plugging used parts in production processes. Data procurement and data analytics is the key to successful implementation of closed loop manufacturing processes. Besides the substantial financial and environmental benefits, industrial sustainability emerges as promising solution to alters the fabrication of global market with an added attention towards responsible manufacturing. Advanced robotics and 3D printing solutions in large scale advanced industries can increase yield, eliminate errors and improve the product life time. As per ongoing study, the "green technology and sustainability market" is not yet reported to be affected due to pandemic outbreak. Infact, green technology and sustainability market can be treated as a new window to position green manufacturing as a catalyst that improves brand image, pushes the industry to seek growth via partnership and innovation in establishing CBM. COVID-19 outbreak brings along economic slowdown, trade disruption, business closures, unemployment, decline in investments and bankruptcies. All of these have a spiral effect on industrial output and growth. The adverse effect of pandemic is very likely to slow down the momentum of sustainable agenda in developing countries. Government need to geostrategies and collaborate to set smaller goals and set new directions for industries to attain sustainable and achievable targets. The challenge in implementing digitization and CBM is opposed by two forces that work against each other – the need to develop resilience and agility to cope with crisis, against the financial constraints (cash preservation). The COVID-19 outbreak has called for a change in the working norms, eliminating the need for noncritical employees to leave their homes, has become a necessity. Digitalization has enabled remote working and collaborations, improving employee safety and operational endurance. Location tracing mobile apps, advanced solutions, such as machine-based algorithms and wearable technologies, have helped in maintaining safe distancing as manufacturing operations restart.

Chapter Summary

In the fourth phase of the revolution, industrial development has assumed a greater role in economic development in any country. The industrial corridors need to assure that the risks associated with industrial development are minimized. Control on water usage and pollution, regulated energy consumption, and investment in renewable energy, elimination of overuse of natural resources, zero-defect in tandem with zero wastage are essential parameters that large-scale industries need to integrate with existing KPI. The impact of Industry 4.0 and its related technology: IIOT, Cyber-physical systems, big data analytics and others, on industrial profitability has now been established and digital transformation is the key to beat competitive disadvantage. On the other hand, the imposition of the non-negotiable success parameter of the 3P (People, Planet and Profit) also referred to as 3E (Environment, Equity and Economics) by governments' worldwide, makes the alliance of technology and sustainability the most feasible evolution path. Research and development, innovation, investment and revenue growth built on zero defect – zero wastage and closed-loop manufacturing systems model will ensure that an ideal eco system thrives where suppliers, manufacturers, employees and consumers pave the foundation for a green economy for future generations.

REFERENCES

Abramov, V.L., F.A. Kodirov, A.A. Gibadullin, V.N. Nezamaikin, O.I. Borisov, and N.V. Lapenkova. (2020) "Formation of Mechanisms for Ensuring the Sustainability of Industry." *Journal of Physics: Conference Series*, 1515(3), p. 3. https://doi.org/10.1 088/1742-6596/1515/3/032025.

Arena, Marika, Natalia Duque Ciceri, Sergio Terzi, Irene Bengo, Giovanni Azzone, and Marco Garetti. (2009) "A State-of-the-Art of Industrial Sustainability: Definitions, Tools and Metrics." *International Journal of Product Lifecycle Management*, 4(1–3), pp. 207–251. https://doi.org/10.1504/IJPLM.2009.031674.

Bandyopadhyay, Santanu, Dominic C.Y. Foo, and Raymond R. Tan. (2017) "Pursuing Sustainability with Process Integration and Optimization." *Process Integration and Optimization for Sustainability*, 1(1), pp. 1–2. https://doi.org/10.1007/s41660-017-0006-1.

Barrutia, Jose M., Carmen Echebarria, Mario R. Paredes, Patrick Hartmann, and Vanessa Apaolaza. (2015) "From Rio to Rio+20: Twenty Years of Participatory, Long Term Oriented and Monitored Local Planning?" *Journal of Cleaner Production*, 106, pp. 594–607. https://doi.org/10.1016/j.jclepro.2014.12.085.

Bilgen, S. (2014) "Structure and Environmental Impact of Global Energy Consumption." *Renewable and Sustainable Energy Reviews*, 38, pp. 890–902. https://doi.org/10.1016/ j.rser.2014.07.004.

Burton, G. Allen, Michelle L. Hudson, Philippa Huntsman, Richard F. Carbonaro, Kevin J. Rader, Hugo Waeterschoot, Stijn Baken, and Emily Garman. (2019) "Weight-of-Evidence Approach for Assessing Removal of Metals from the Water Column for Chronic Environmental Hazard Classification." *Environmental Toxicology and Chemistry*, 38(9), pp. 1839–1849. https://doi.org/10.1002/etc.4470.

Colucci, Mariachiara, Annamaria Tuan, and Marco Visentin. (2020) "An Empirical Investigation of the Drivers of CSR Talk and Walk in the Fashion Industry." *Journal of Cleaner Production*, 248, pp. 119200. https://doi.org/10.1016/j.jclepro.2019.119200.

Demartini, Melissa, Steve Evans, and Flavio Tonelli. (2019) "Digitalization Technologies for Industrial Sustainability." *Procedia Manufacturing* 33, pp. 264–271. https://doi.org/1 0.1016/j.promfg.2019.04.032.

Fröhling, Magnus, and Michael Hiete. (2020) "Sustainability and Life Cycle Assessment in Industrial Biotechnology: A Review of Current Approaches and Future Needs." *Advances in Biochemical Engineering/Biotechnology*, 173, pp. 143–203. https:// doi.org/10.1007/10_2020_122.

Grappi, Silvia, Simona Romani, and Camilla Barbarossa. (2017) "Fashion without Pollution: How Consumers Evaluate Brands after an NGO Campaign Aimed at Reducing Toxic Chemicals in the Fashion Industry." *Journal of Cleaner Production* 149, pp. 1164–1173. https://doi.org/10.1016/j.jclepro.2017.02.183.

Holden, Erling, Kristin Linnerud, and David Banister. (2014) "Sustainable Development: Our Common Future Revisited." *Global Environmental Change*, 26(1), pp. 130–139. https://doi.org/10.1016/j.gloenvcha.2014.04.006.

Ii, Part. (2012) "Indian Minerals Yearbook 2011." *Indian Minerals*, 2011 (October), pp. 1–9.

Kumar, Aalok, Sunil Luthra, Sachin Kumar Mangla, and Yiğit Kazançoğlu. (2020) "COVID-19 Impact on Sustainable Production and Operations Management." *Sustainable Operations and Computers*, 1(June), pp. 1–7. https://doi.org/10.1016/j.susoc.2020.06.001.

Le, Jennifer T., Lisa A. Levin, and Richard T. Carson. (2017) "Incorporating Ecosystem Services into Environmental Management of Deep-Seabed Mining." *Deep-Sea Research Part II: Topical Studies in Oceanography*, 137, pp. 486–503. https://doi.org/10.1016/j.dsr2.2016.08.007.

Lin, Ru-Jen, Rong-Huei Chen, and Thi-Hang Nguyen. (2011) "Green Supply Chain Management Performance in Automobile Manufacturing Industry under Uncertainty." *Procedia - Social and Behavioral Sciences* 25, pp. 233–245. https://doi.org/10.1016/j.sbspro.2011.10.544.

Ludo, Alcorta. (2017) "Delivering the Sustainable Development Goals," 40.

Mascarenhas, Jefferson dos Santos, Hemal Chowdhury, M. Thirugnanasambandam, Tamal Chowdhury, and R. Saidur. (2019) "Energy, Exergy, Sustainability, and Emission Analysis of Industrial Air Compressors." *Journal of Cleaner Production* 231, pp. 183–195. https://doi.org/10.1016/j.jclepro.2019.05.158.

Ng, Tan Ching, and Morteza Ghobakhloo. (2020) "Energy Sustainability and Industry 4.0." *IOP Conference Series: Earth and Environmental Science* 463(1), pp. 2–3. https://doi.org/10.1088/1755-1315/463/1/012090.

Parida, Vinit, David Sjödin, and Wiebke Reim. (2019). "Reviewing Literature on Digitalization, Business Model Innovation, and Sustainable Industry: Past Achievements and Future Promises." *Sustainability (Switzerland)*, 11(2), p. 9. https://doi.org/10.3390/su11020391.

Pereira, A.C., and F. Romero. (2017) "A Review of the Meanings and the Implications of the Industry 4.0 Concept." *Procedia Manufacturing*, 13, pp. 1206–1214. https://doi.org/10.1016/j.promfg.2017.09.032.

Sandstrom, Gregory. (2020) "Editorial: Insight." *Technology Innovation Management Review* 10(2), pp. 3–4. https://doi.org/10.22215/timreview/1323.

Spoelstra, Sierk F. (2013). "Sustainability Research: Organizational Challenge for Intermediary Research Institutes." *NJAS - Wageningen Journal of Life Sciences*, 66, pp. 75–81. https://doi.org/10.1016/j.njas.2013.06.002.

Stock, T., and G. Seliger. (2016) "Opportunities of Sustainable Manufacturing in Industry 4.0." *Procedia CIRP*, 40(Icc), pp. 536–541. https://doi.org/10.1016/j.procir.2016.01.129.

Sutherland, John W., Steven J. Skerlos, Karl R. Haapala, Daniel Cooper, Fu Zhao, and Aihua Huang. (2020) "Industrial Sustainability: Reviewing the Past and Envisioning the Future." *Journal of Manufacturing Science and Engineering*, 142(11), pp. 1–16. https://doi.org/10.1115/1.4047620.

Szolnoki, Gergely, Dieter Hoffmann, Pricewater Coopers, Price Waterhouse Cooper, John Overton, Warwick E. Murray, Alecia Boshoff, Gergely Szolnoki, B. Morokolo, and United Nations Global Compact KPMG. (2014) "SDG Industry Matrix." *United Nations* 53 (December), pp. 1–48. https://www.unglobalcompact.org/docs/issues_doc/development/SDGMatrix-ConsumerGoods.pdf%5Cn http://dx.doi.org/10.1016/j.jclepro.2013.03.045 %5Cn http://dx.doi.org/10.1108/IJWBR-10-2012-0028%5Cn; http://dx.doi.org/10.1108/IJWBR-04-2013-0015%5Cn http://dx.doi.org.

Tonelli, Flavio, Steve Evans, and Paolo Taticchi. (2013) "Industrial Sustainability: Challenges, Perspectives, Actions." *International Journal of Business Innovation and Research*, 7(2), pp. 143–163. https://doi.org/10.1504/IJBIR.2013.052576.

United Nations Global Compact & KPMG. (2017) "SDG Industry Matrix," 58. https://home.kpmg.com/content/dam/kpmg/xx/pdf/2017/05/sdg-energy.pdf.

Xavier, Lúcia, Helena Ellen, Cristine Giese, Ana Cristina, Ribeiro Duthie, and Fernando Antonio, and Freitas Lins. (Aug. 2019) "Sustainability and the Circular Economy: A Theoretical Approach Focused on e-Waste Urban Mining." *Resources Policy*, Art. no. 101467. https://doi.org/10.1016/j.resourpol.2019.101467.

Zaccai, Edwin. (2012) "Over Two Decades in Pursuit of Sustainable Development: Influence, Transformations, Limits." *Environmental Development*, 1(1), pp. 79–90. https://doi.org/10.1016/j.envdev.2011.11.002.

Web References

Agrawal, Mayank, Karel Eloot, Matteo Mancini, and Alpesh Patel. (Jun. 2020) "Industry 4.0: Reimagining Manufacturing Operations after COVID-19." https://www.mckinsey.com/business-functions/operations/our-insights/industry-40-reimagining-manufacturing-operations-after-covid-19

Industrial Development Report 2018, "Demand for Manufacturing: Driving Inclusive and Sustainable Industrial Development." https://sustainabledevelopment.un.org/content/documents/2537IDR2018_FULL_REPORT_1.pdf

White et al. (2019) "The Elusive Green Consumer." https://hbr.org/2019/07/the-elusive-green-consumer

3 Sustainable Supplier Selection in Industry 4.0: A Three-Stage Fuzzy Kano and FIS-Based Decision Framework

Naveen Jain[1] and A. R. Singh[2]
[1]Shri Shankracharya Institute of Professional Management and Technology, Raipur, India
[2]National Institute of Technology Raipur, Raipur, India

3.1 INTRODUCTION

Today, we all are in the mid of the transformation era where products are manufactured through digitized manufacturing processes. This transformation is called Industry 4.0 (I40) and has been considered the fourth industrial revolution by experts. I40 emphasizes connecting computers to communicate and to finally make a decision without human involvement. It is only possible by combining the four design principles of I40: interconnection, real-time information transparency, decentralization, technical assistance (Hermann et al., 2016). Cyber-Physical Systems (CPS), the Internet of Things (IoT), and the Internet of Systems make I40 the most promising future technology. Cyber-physical systems use modern control systems having embedded software systems and are managed via IoT. Hence, products and machines are integrated and can "communicate" with each other. This is a significant departure from the traditional manufacturing ways and provides a definite way for making smart factories a reality. With the replacement of traditional manufacturing practices by smart and efficient technologies, the productivity of the industries has reached new heights. Industries can now manufacture high-quality products on a large scale in the minimum possible time. This has resulted in the consumption and procurement of large quantities of raw materials by industries, especially large-scale industries. This procurement and consumption of a high amount of raw materials has raised a serious concern about the adverse environmental impact of industrial activity. Further, the growing awareness among the various stakeholders, stringent government laws, and vigilant Government/ NGOs have compelled industries to give serious thought to the application and integration

DOI: 10.1201/9781003102304-3

of sustainability aspects in their supply chains. Further, the I40 manufacturing process has multiple and far-reaching effects on the entire sustainability aspects of SC of the industry. I40 significantly affects all three dimensions of the TBL approach. I40 demands huge capital investment and also increases the production rate of the industry, which, in turn, requires the ready availability of a huge amount of resources, like minerals and raw materials. Procuring large quantities of raw materials does not only require more capital but also demands more excavation and clearing of forests, thus putting extra pressure on the environmental aspect of sustainability. I40 has also been a centre point of discussion from the social dimension of sustainability as it is held responsible for cutting jobs. Hence, close cooperation and coordination among SC members is vital for increasing the information visibility in various stages of the product life cycle for efficient establishment of sustainability in the SC.

Suppliers, as key members of the upstream supply chain, significantly help industries achieve their sustainability of objectives in SC. Supply chain decisions are very critical and strategic for industry management for the implementation of Sustainable Supply Chain (SSC). The industry image and competitive position are largely influenced by the suppliers the industry has. Researchers and academicians have addressed the problem and proposed many Sustainable Supplier Selection (SSS) methodologies (Zimmer et al., 2016). While great efforts have been made by researchers in developing these methodologies, little attention has been given to incorporating the TBL (Triple Bottom Line) dimensions (economic, social, and environmental) within the context of I40 SC and sustainable supplier selection evaluation and selection, specifically. Further, the incorporation of large criteria under all three TBL dimensions for SSS poses a greater challenge to researchers for developing an efficient decision framework. The Kano model is popular for criteria clustering because it provides an easy platform to measure the perceived level of satisfaction of DM against the product attributes (Jain & Singh, 2019). The model helps identify the most essential and elementary SSS criteria, which need to be incorporated for developing a robust SSS process. In the SSS process, the industry acts as a customer for products supplied by the suppliers. Hence, in this work, FKM has been applied for clustering of evaluation criteria in each sustainability dimension into various useful clusters, like most essential (must-be), essential (one dimensional), desirable (attractive), and not affecting (indifferent). As it is known, the Kano model is a questionnaire-based technique that captures the responses from DMs for criteria clustering. As it is well known that ambiguity and uncertainty are generally inherited in all human responses, there is a strong necessity to capture this uncertainty in responses by integrating fuzzy logic with the Kano model.

It elicits from the literature that most of the sustainable supplier-selection decision frameworks developed by academicians and researchers are MCDM based. In the MCDM technique based methodology, DMs need to deal with linguistic responses to evaluate a supplier's performance. However, the subjectivity of response is missed as qualitative perceptions are transferred to quantitative terms (Amindoust et al., 2012). To overcome this challenge, some of the academicians have integrated fuzzy logic with MCDM. However, the integration of fuzzy logic with MCDM techniques does not facilitate the challenge of handling the enormous number of sustainability criteria, and when the number of alternatives is more, the mathematical calculations become complex.

To overcome these challenges of capturing DMs' ambiguity in responses and having a methodology that can handle large sustainability criteria as well as a large number of alternatives for selection, this work applies the Fuzzy Inference System (FIS) for SSS in the context of I40. The supplier's performance is evaluated in FIS integrating the knowledge of DMs for framing the fuzzy rules for decision making. This leads to precise and effective supplier evaluation, leading to more reliable and accurate results. Hence, using a FIS-based methodology, DMs need to take optimum decisions regarding the number of rules and the desired accuracy.

The proposed methodology has three stages. In the first stage, the comprehensively established sustainability criteria in all three TBL dimensions are categorized into different clusters as per their priority within the FKM. In the second stage, preselection of the potential suppliers is performed by evaluating their performance against the most essential (Must be) criteria. In the third and final stage, the suppliers qualifying the pre-selection stage are assessed against the desirable (Attractive) sustainability criteria for the final selection of suppliers.

The proposed work contributes to the technical community in the domain of sustainable supplier selection in the context of I40 in the following ways: (i) Identification of most essential and desirable sustainability criteria for SSS; (ii) Making three modified architecture-enabled FIS engines that help DMs frame least number of rules for each engine in each TBL dimension; (iii) Finally, the SSS selection framework facilitates evaluating Supplier Performance Index (SPI) for all suppliers in three TBL dimensions and based on SPI values, the final selection is made by decision-makers.

The rest of the article is organized as follows. In section 2, a literature review on Industry 4.0, Sustainable Supplier Selection (SSS) in Industry 4.0, Fuzzy Kano Model (FKM), and Fuzzy Inference System (FIS) has been provided. Section 3 presents the proposed FKM-based FIS model. Section 4 covers the application of the proposed model and results. Section 5 covers the Managerial applications of the proposed work. Section 6 covers the conclusion and future scope of work.

3.2 LITERATURE REVIEW

3.2.1 INDUSTRY 4.0

I40 has specifically redefined the traditional manufacturing process and has given a totally new dimensional approach to the way production shop floors should be operated for increased productivity. Many experts have defined I40 as a global transformation of the manufacturing industries aided by tools like digitalization and the Internet of things. Although I40 originated in Germany, it shares commonalities with European countries, where it is known as Smart Factories, Smart Industry, Advanced Manufacturing, or Industrial Internet of Things (IIoT) (Zhong et al., 2017). Considered as the fourth industrial revolution, it has increased the operational effectiveness of industries along with the development of newer business models, products, and services. I40 aims at innovating the manufacturing processes of the industries, thus bringing an improvement in the sustainability achievement status of the supply chains. I40 can be regarded as a mix of leading-edge technologies to innovate the assembly lines of industries (Thoben et al., 2017; Herrmann et al., 2014).

- *Internet of Things(IoT):-* The Internet of Things (IoT) refers to the wide network of physical objects—"things"—equipped with sensors, software, and other technologies to connect and exchange data with each other using the internet. These devices range from ordinary household appliances such as kitchen appliances, cars, thermostats, baby monitors to sophisticated industrial tools.
- *Robotics:-*Robots are machines that resemble human beings and are capable of replicate certain human movements and gestures automatically. Robots are programmed to perform many industrial functions and tasks such as modelling, welding, gripping, spinning, fastening, etc. Robots make them themselves vital due to their capability to do various tasks related to data analysis thus providing insights for interpretation and action by the management and decision-makers (Demir et al., 2019).
- *Big Data Analytics:* I40 generates a large amount of unstructured data from Product and/or machine design data, Machine-operation data, Product- and process-quality data, Fault-detection, and other system-monitoring deployments, etc. These data are stored in central servers with the help of cloud computing technologies and need to be analyzed for extracting useful information by applying machine learning and data analytics techniques (Yin & Kaynak, 2015).
- *Cloud Computing:* It is the delivery of on-demand computing services varying from applications, storage, processing power over the internet. It helps the organizations with IT facilities and helps organizations to perform economically (Xu, 2012).
- *Additive Manufacturing (AM):-* AM or Additive layer manufacturing (ALM) is the industrial production name for 3D printing.

I40 integrates and applies these technologies to the assembly line and build CPS where real-time data about the production process is generated. Further, this data is collected and analyzed and made available to the management for decision making and future planning (Braccini & Margherita, 2018). Further, these technologies help organizations develop advanced manufacturing solutions, integrating machines, and thus offering greater production flexibility and enhanced quality and performance.

Researchers have stressed I40 as a facilitator for organizations in establishing and enhancing the sustainability aspects under TBL dimensions. Many scholars believe that I40 helps ease down the hard muscular work of labourers by the installation of smart and intelligent machines and robots (Rajnai & Kocsis, 2017). I40 also creates new sustainable markets for industries, helps in shortening the lead times of products, and helps in increasing the production rate and flexibility, thus enhancing the efficient use of available resources (Lasi et al., 2014; Waibel et al., 2017). However, on the other hand, some academicians hold a negative opinion about I40 (Horváth & Szabó, 2019). I40 empowers industries with modern, superior, and smarter technologies and machines, thus enabling them with a higher production rate, which results in more consumption of natural resources and ultimately affects the environment (Yang et al., 2018; Stock & Seliger, 2016; Müller et al., 2018). Further, I40 employs robots, smart machines, and storing huge amounts of data on cloud servers, which demands more electrical power thus resulting in the consumption of environmental resources at a faster rate and increasing pollution (Kamble et al., 2018). I40 also impacts the social dimension of the TBL approach. Researchers fear that as robots are fast replacing the human workforce, soon humankind will be faced with challenges of jobs and livelihood (Kiel et al., 2017).

3.2.2 SUSTAINABLE SUPPLIER SELECTION FOR INDUSTRY 4.0

I40 has redefined the whole supply chain structure of industries by introducing digitization and automation of all processes. Today's SC consists of a network of CPS communicating through IoT and exchanging vital information among them in real-time (Ghadimi et al., 2019). Having undergone a face shift, the activities of SC need to be aligned in the context of the sustainability dimension also. Researchers have addressed the need of achieving sustainability in I40 and the action to be taken for it. Müller et al. (2018) through their work have highlighted the opportunities and challenges for the implementation of I40 in the context of sustainability. Various other researchers have addressed the methods of integrating the sustainability initiatives in SC of the industry through I40 (Hermann et al., 2016; Bonilla et al., 2018; Gabriel & Pessl, 2016).

However, research on SSS and FKM in the context of I40 is at an early stage. Hence, research into incorporating sustainability aspects in supplier performance evaluation and selection to meet the I40 SC sustainability needs more erudite attention and practical demonstrations. Moreover, the reviewed studies that addressed the process of supplier selection using an agent-based approach, considered the single-stage (either pre-selection or final selection) decision framework only. Based on the above-identified gaps, the theoretical underpinnings of the proposed work distinguish itself from previous literature related to the SSS by proposing a three-stage FKM-based FIS approach.

3.2.3 FUZZY KANO MODEL (FKM)

Kano et al. (1984) adapted the "Motivation-Hygiene Theory" and proposed a two-way model on quality, known as the theory of attractive quality. Kano's model is based on the customer's experience and perception and focuses on the most important attributes of the product that contribute to increasing the customer satisfaction level. By having an understanding of customer satisfaction and product attributes, decision-makers can have products aligned with requirements and can achieve higher customer satisfaction levels. The Kano model is a questionnaire-based method with a pair of questions for each criterion. Respondents have to mark a single response for each question, out of five available choices. However,

TABLE 3.1

Domains Fuzzy Kano Model Application

S.No.	Author	Domain
1	Jain et al. (2016)	Supplier selection
2	Wang and Wang (2014)	Smart cameras
3	Lee et al. (2008)	Product life cycle management
4	Florez-Lopez and Ramon-Jeronimo (2012)	Customer service logistics
5	Chen and Ko (2008)	Semiconductor packing
6	Vinodh et al. (2013)	Sustainability assessment in industry
7	Hannan et al. (2015)	Examination management software
8	Pai et al. (2018)	Chain restaurant industry
9	Wang and Fong (2016)	Airline industry
10	Shafia and Abdollahzadeh (2014)	Product coding system
11	Wang (2013)	New smart pads
12	Chyu and Fang (2014)	New product development
13	Naveen Jain and Singh (2020)	Sustainable supplier selection

according to Manski (1990), human judgment is accompanied by uncertainty and because of this traditional, questionnaires are interpreted wrongly. Further, Huang and Wu (1992) reported that customers' responses cannot be truly captured by a single response in a questionnaire. To capture uncertainty and vagueness and to have multiple responses from the respondent, Lee and Huang (2009) proposed the fuzzy Kano model. The FKM lets respondents mark more than one response for each question, thus providing a better understanding of the satisfaction perceived by them. The various domains where the researchers have successfully applied the FKM (Table 3.1)

The FKM classifies the criteria into following categories.

i) *Must-be requirements (M)*, ii) *One-dimensional requirements* (O) iii) *Attractive requirements (A)*: iv) *In different requirements (I)*: v) *Reverse requirements (R)*:

3.2.3.1   Steps for Fuzzy Kano Model Application

Step 1. Identification of criteria/attributes to be classified.
Step 2. Preparation of Fuzzy Kano Questionnaire (FKQ).
Step 3. Analysis of the functional and dysfunctional forms of questions.
Step 4. Establishment of 5 x 5 relation matrix using relation

$$R = F' \ x \ D$$

Where R is the fuzzy relation matrix

TABLE 3.2
Kano Evaluation Table

The functional form	Dysfunctional form				
	Like	Must Be	Neutral	Live with	Dislike
Like	Questionable	Attractive	Attractive	Attractive	One dimensional
Must Be	Reversible	Indifferent	Indifferent	Indifferent	Must be
Neutral	Reversible	Indifferent	Indifferent	Indifferent	Must be
Live with	Reversible	Indifferent	Indifferent	Indifferent	Must be
Dislike	Reversible	Reversible	Reversible	Reversible	Questionable

F' is the transpose of the functional vector-matrix

D is the dysfunctional vector-matrix

Step 5. Comparison of matrix R with the Kano evaluation table (Table 3.2).

Step 6. Summing all values corresponding to the attribute and finalization of the Kano category.

3.2.4 Fuzzy Logic and Fuzzy Inference System

The word "Fuzzy" refers to things that are imprecise/vague/ambiguous. Fuzzy logic is a method based on logic operations and considers many-valued logic rather than binary logic. Modern computers work on "true or false" (0 to 1), Boolean logic, and can handle data that represents subjective or vague ideas. Thus, Fuzzy logic provides an advantage to decision-makers by providing truth values between 0 and 1, and these values are designated as intensity (degree) of truth.

3.2.4.1 Fuzzy Logic

Fuzzy logic allows computers to handle imprecise data. The characteristics which make fuzzy logic the preferred choice for decision-makers are:

- Fuzzy Logic is flexible and easy to implement.
- It helps in capturing the logic of human thought.
- It has a multi-valued output that represents multiple possible solutions.
- Highly suitable method for capturing uncertainty or approximate reasoning.
- Fuzzy logic incorporates the knowledge of experts, which helps achieve better results.

3.2.4.2 Fuzzy Inference System (FIS)

A FIS helps in mapping of input with output using fuzzy logic. The mapping helps DMs make the appropriate decision. The fuzzy inference system incorporates membership functions, fuzzy logic operators, and rules.

The working of the FIS consists of the following steps (Figure 3.1):

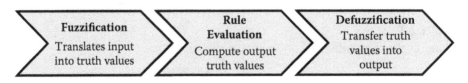

FIGURE 3.1 FIS system

1. **Fuzzification:** - A fuzzification unit supports the application of fuzzification methods and converts the crisp input into fuzzy input.
2. **Rule Evaluation:** - a collection of rule base and database is formed upon the conversion of crisp input into fuzzy input.
3. **Defuzzification:**-The defuzzification unit converts the fuzzy output into crisp output.

The proposed FIS model development consists of three major steps:

a. *Membership function and operational operators*

In this step, every sustainability criteria finalized is considered and based on DMs' experience, the evaluation levels are established. As the evaluation levels are established on experience, they are subjective and there is a need to convert them into the membership grade. In both the pre-selection and final stage, the triangular membership function for the input variables is considered. A fuzzy number in a triangular form can be represented as $\widetilde{M}$ = (a, b, c), as shown in the Figure. 3.2 and defined as Equation (3.1).

$$
\mu_{\tilde{M}} = \begin{cases} 0, & x < a, \\ \frac{(x-a)}{(b-a)}, & a \le x \le b \\ \frac{(c-x)}{(c-x)}, & b \le x \le c \\ 0, & x > c, \end{cases} \tag{3.1}
$$

FIS models apply to inputs as "Best" "Worst" and "Average" as linguistic terms used by decision-makers for evaluating supplier's performance (Table 3.3) (Figure 3.2).

b. *Fuzzy rules (KM based) and Modified Architecture of FIS engine*

In this step, the fuzzy rules are framed employing the expert's knowledge for FIS models. The number of fuzzy rules in a FIS model is determined by using Equation (3.2).

$$
\textit{No. of fuzzy rules} = (\textit{Membership functions})^{\textit{No. of input variables}} \tag{3.2}
$$

TABLE 3.3

Linguistic Terms for Supplier Performance Evaluation

Linguistic terms used by DM's	Fuzzy set
Best	(b, c, c)
Average	(a, b, c)
Worst	(a, a, b)

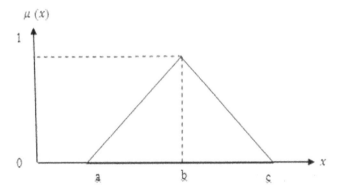

FIGURE 3.2 The triangular fuzzy membership function

c. *Defuzzification*

The output of the FIS model is further defuzzified in this step. Defuzzification is done to get the results in the form of real numbers. In this work, the Center Of Area (COA) method has been used for the defuzzification of the fuzzified results and is expressed mathematically as Equation (3.3)

$$x_{COA} = \frac{\sum_{i=1}^{n} x_i \mu_i(x_i)}{\sum_{i=1}^{n} \mu_i(x_i)} \tag{3.3}$$

The defuzzification process results in the determination of the Economic Performance Index (EOPI), Social Performance Index (SOPI), and the Environmental Performance Index (ENPI) for each potential supplier. EOPI, SOPI and ENPI are used collectively to determine the final Sustainability Performance Index (SPI) for the suppliers. SPI is the summation of EOPI, SOPI, and ENPI. Mathematically

$$SPI = \sum \{ (W_{economic} x EOPI) + (W_{social} x SOPI) + (W_{environmental} x ENPI) \} \tag{3.4}$$

$W_{economic}$, W_{social} and $W_{environmental}$ are the importance of weight respective sustainability dimensions. The values of the importance of weight can vary from 0 to 1 as per the need of the case industry. High priority is indicated by value 1 and vice versa. The value of SPI at the end of the pre-selection and final selection stage facilitates DMs final selection of suppliers.

3.3 PROPOSED METHODOLOGY

To accomplish the research objectives, a three-stage FKM integrated FIS model has been developed, which incorporates Must be and Attractive sustainability criteria in three TBL dimensions. The steps of the decision framework are shown in Figure 3.3.

Step 1: Formation of Decision-makers' team

Initially, a team comprising of industry experts selected from middle/ top management is to be constituted. The responsibility of the team is to identify and constitute a pool of suppliers from which the final selection can be made. Secondly, the team will be performing the task of finalizing the sustainability criteria for supplier evaluation. Thirdly, the team will assess the suppliers and will provide a linguistic response against the supplier's evaluation. Finally, the team will be responsible for framing the decision rules for the FIS system.
Step 2: Identification of sustainability criteria and application of fuzzy Kanomodel for clustering

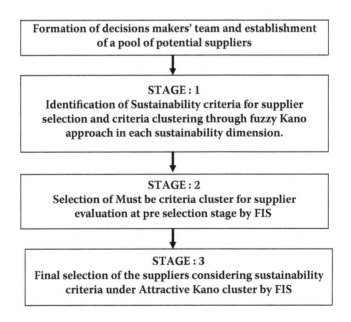

FIGURE 3.3 Flow chart of the proposed methodology

Literature review is to be performed for the establishment of the sustainability criteria for SS. Further FKM will be applied for clustering the sustainability criteria into various Kano categories.

Step 3: FIS model for pre-selection of suppliers considering Must be sustainability criteria

In this step, sustainability criteria under the Must-be Kano category are considered for pre-selection of the potential suppliers. The proposed FIS model has three distinct FIS engines considering Must be sustainability criteria in each TBL dimension.

Step 4: FIS model considering Attractive Kano criteria

Here the final selection is done by considering sustainability criteria under the Attractive Kano category. A FIS model comprising of three engines needs to be developed catering to each sustainability dimension for final evaluation and selection.

3.4 APPLICATION OF PROPOSED METHODOLOGY

The Indian iron and steel sector has contributed significantly to achieving a strong GDP and helped in securing the seventh position worldwide with a GDP of $2.72 trillion. Fast development in the core sectors such as railways, infrastructure, airports, oil & gas, and armed forces has made the steel industry the backbone of national growth. According to the report of World steel, in 2018, India produced 106.5 MT of crude steel 106.5 MT, up 4.9% from 101.5 MT in 2017. Thus, India became the world's second-largest steel producing country.

Indian iron and steel industry caters to the need of customers around the world and manufacture products like rails, plates, bloom, billets, wheel, wires, rods and axles, etc. India strives hard to become the worldwide leader in crude steel production. India plans to achieve this by incorporating I40 practices to enhance steel production manifolds. However, the steel ministry also aims to achieve sustainability in SC. Hence the iron and steel industry has been considered as a case study in this work. The flowchart of the proposed work is shown in Figure 3.3:

Step 1: Formation of Decision-makers' team

As a prerequisite of the proposed methodology, a team of decision-makers comprising of four members is made. The team members were identified to represent the management of the industry and belong to the departments connected with the supplier selection process. A prior meeting was held with the team and the team members were briefed about the nature of responsibilities they ought to carry out. DM details are shown in Table 3.4. The potential suppliers considered for the methodology are seven in number.

TABLE 3.4
Details of Decision Makers

	Department	Experience in years	Designation
DM 1	Materials management	13	Senior manager
DM 2	Finance Department	21	Deputy general manager
DM 3	Stores management	17	Assistant general manager
DM 4	Quality control	9	Manager

The sustainability criteria for evaluation in three TBL dimensions are shown in Table 3.5, 3.6 and 3.7

Step 2: Identification of sustainability criteria and application of the FKM for clustering

In this step, three separate FKQ, one each for all three TBL dimensions were established based on the sustainability criteria as depicted in step 1. The total number of one hundred thirty-two questions were framed in which thirty-six question comprises for economic criteria, FKQ for social criteria comprises thirty-four, and FKQ for environmental dimension criteria contained thirty-four questions. The DMs were presented with these three FKQ. DMs gave their responses ranging from 0% to 100%. The responses were screened for valid responses. A sample for Quality criteria FKQ is shown in Table 3.8. The FK model was applied to all the sustainability criteria and the Kano clustering has been established as shown in Table 3.9.

Step 3: FIS model for pre-selection of suppliers considering Must be sustainability criteria

In this step, FIs engines are developed considering the Must be Criteria for the pre-selection of the sustainable supplier. Three separate FIS engines are established for all sustainability dimensions (Fig 3.4). Must be criteria are the criteria that every supplier is expected to fulfil for the pre-selection stage (Figure 3.4).

a. **Membership function and operational operators**

Three linguistic rating variables used are "Best", "Average", and "Worst", as shown in Figure 3.5. The three levels for quality criteria have been shown in Table 3.10. The membership functions for the evaluation of EOPI, ENPI and the SOPI and linguistic variables used are shown in Figure 3.6

TABLE 3.5

Sustainability Criteria in Economic Dimension for Supplier Evaluation

S.No.	Economic criteria	Authors
1	Quality • (Quality Performance) • (Quality control) • (Quality management) • (Quality Certification) • Quality assurance ability)	Lima Junior et al. (2014); Kar (2014); Guo et al. (2014); Ware et al. (2014); Deng et al. (2014); Osiro et al. (2014); Qian (2014); Rouyendegh and Saputro (2014); Kannan et al. (2014); Rajesh and Malliga (2013); Shen and Yu (2013); Junior et al. (2013); García et al. (2013); Omurca (2013); Mukherjee and Kar (2013); Kannan et al. (2013); Arikan (2013); Ekici (2013); Kumar et al. (2013); Chen and Wu (2013); Nazari-Shirkouhi et al. (2013); Dey et al. (2012); Azadnia et al., 2012); Ghorbani et al. (2012); Peng (2012); Amindoust et al. (2012); Khaleie et al. (2012); Erdem and Göçen (2012); Zouggari and Benyoucef (2012); Xiao et al. (2012); Liao and Kao (2011); Kilincci and Onal (2011); Aksoy and Öztürk (2011); Lam et al. (2010); Sanayei et al. (2010); Bai and Sarkis (2010); Che and Wang (2010); Liao and Kao (2010); Wang (2010); Emre et al. (2009); Shen and Yu (2009); Montazer et al. (2009); Guo et al. (2009); Ayhan and Kilic (2015); Orji and Wei (2015); Karsak and Dursun (2015); Lee et al. (2015); Akman (2015); Beikkhakhian et al. (2015); Wu et al. (2016); Galankashi et al. (2016); Huang and Keskar(2007); Dweiri et al. (2016); Ahmadi et al. (2017); Chaharsooghi and Ashrafi (2014); Fallahpour et al. (2017); Luthra et al. (2017); Rao et al. (2017); Kaur et al. (2016); Kannan (2018); Ghadimi et al. (2017); Zimmer et al. (2016)
2	Delivery • (Lead Time) • (Time to Deliver) • (Responsiveness) • (Delivery Performance)	Lima Junior et al. (2014); Kar (2014); Guo et al. (2014); Ware et al. (2014); Osiro et al. (2014); Jadidi et al. (2014); Qian (2014); Rouyendegh and Saputro (2014); Kannan et al. (2014); Guo et al. (2014); Rajesh and Malliga (2013); Junior et al. (2013); García et al. (2013); Omurca (2013); Mukherjee and Kar (2013); Kannan et al. (2013); Kumar et al. (2013); Chen and Wu (2013); Dey et al. (2012); Azadnia et al. (2012); Ghorbani et al.(2012); Amindoust et al. (2012); Khaleie et al. (2012); Erdem and Göçen (2012); Zouggari and Benyoucef (2012); Hsu et al. (2012); Xiao et al. (2012); Kilincci and Onal (2011); Aksoy and Öztürk (2011); Dogan and Aydin (2011); Sanayei et al. (2010); Bai and Sarkis (2010); Che and Wang (2010); Liao and Kao (2010); Wang (2010); Emre et al. (2009); Shen and Yu (2009); Montazer et al. (2009); Guneri et al. (2009); Guo et al. (2009); Ayhan and Kilic (2015); Lee et al. (2015); Akman (2015); Beikkhakhian et al. (2015); Dweiri et al. (2016); Lima-Junior and Carpinetti,(2016); Chaharsooghi and Ashrafi (2014); Fallahpour et al. (2017);

(Continued)

TABLE 3.5 (Continued)
Sustainability Criteria in Economic Dimension for Supplier Evaluation

S.No.	Economic criteria	Authors
		Luthra et al. (2017); Rao et al. (2017); Kaur et al. (2016); Kannan (2018)
3	Performance history	Deng et al. (2014); Xiao et al. (2012); Lam et al. (2010); Wang (2010); Hassanzadeh and Razmi (2009); Guo et al. (2009); Lee et al. (2015)
4	Warranties and claim policies • (Warranty)	Guo et al. (2014); Liao and Kao (2010); 2011; Guo et al. (2009)
5	Production facilities • (Production capability) • (Capacity)	Jadidi et al. (2014); Rouyendegh and Saputro (2014); Omurca (2013); Ekici (2013); Nazari-Shirkouhi et al. (2013); Xiao et al. (2012); Guo et al. (2009); Lee et al. (2015); Luthra et al. (2017); Kannan (2018); Ghadimi et al. (2017)
6	Net price • (Cost) • (Purchasing Price)	Lima Junior et al. (2014); Aksoy et al. (2014); Guo et al. (2014); Ware et al. (2014); Deng et al. (2014); Osiro et al. (2014); Jadidi et al. (2014); Qian (2014); Rouyendegh and Saputro (2014); Kannan et al. (2014); Guo et al. (2014); Rajesh and Malliga (2013); Shen and Yu (2013); García et al. (2013); Omurca (2013); Mukherjee and Kar (2013); Kannan et al. (2013); Ekici (2013); Chen and Wu (2013); Ruiz-Torres et al. (2013); Nazari-Shirkouhi et al. (2013); Dey et al. (2012); Azadnia et al. (2012); Ghorbani et al. (2012); Peng (2012); Amindoust et al. (2012); Khaleie et al. (2012); Erdem and Göçen (2012); Zouggari and Benyoucef (2012); Aksoy and Öztürk (2011); Dogan and Aydin (2011); Lam et al. (2010); Sanayei et al. (2010); Bai and Sarkis (2010); Che and Wang (2010); Liao and Kao (2010); Wang (2010); Emre et al. (2009); Montazer et al. (2009); Guo et al. (2009); Huang and Keskar (2007); Ayhan and Kilic (2015); Karsak and Dursun (2015); Lee et al. (2015); Akman (2015); Beikkhakhian et al. (2015); Wu et al. (2016); Galankashi et al. (2016); Dweiri et al. (2016); Lima-Junior and Carpinetti (2016); Ahmadi et al. (2017); Fallahpour et al. (2017); Luthra et al. (2017); Rao et al. (2017); Kaur et al. (2016); Zimmer et al. (2016)
7	Technical capability • (Capability) • Technology) • (R & D Adaptability)	Kar (2014); Osiro et al. (2014); Rouyendegh and Saputro (2014); Kannan et al. (2014); Junior et al. (2013); Omurca (2013); Kannan et al. (2013); Peng (2012); Khaleie et al. (2012); Erdem and Göçen (2012); Xiao et al. (2012); Kilincci and Onal (2011); Dogan and Aydin (2011); Sanayei et al. (2010); Bai and Sarkis (2010); Bai and Sarkis (2010); Guo et al. (2009); Lee et al. (2015); Wu et al. (2016); Galankashi et al. (2016); Chaharsooghi and Ashrafi (2014); Luthra et al. (2017); Kaur et al. (2016); Kannan (2018); Ghadimi et al. (2017); Zimmer et al. (2016)

TABLE 3.5 (Continued)
Sustainability Criteria in Economic Dimension for Supplier Evaluation

S.No.	Economic criteria	Authors
8	Financial position • (Financial Health) • (Financial performance) • (Financial strength)	Kar (2014); Osiro et al. (2014); Rouyendegh and Saputro (2014); Ghorbani et al. (2012); Xiao et al. (2012); Kilincci and Onal (2011); Dogan and Aydin (2011); Hassanzadeh and Razmi (2009); Guo et al. (2009); Lee et al. (2015); Wu et al. (2016); Chaharsooghi and Ashrafi (2014); Luthra et al. (2017); Kaur et al. (2016); Kannan (2018)
9	Procedural compliance	Dickson (1966)
10	Communication system • (Ease of communication)	Rouyendegh and Saputro (2014); Junior et al. (2013); Xiao et al. (2012); Dalalah et al. (2011); Hassanzadeh and Razmi (2009); Dogan and Aydin (2011); Ebrahim et al. (2009); Guo et al. (2009); Araz and Ozkarahan (2007)
11	Reputation and position in the industry	Rouyendegh and Saputro (2014); Ghorbani et al. (2012); Zouggari and Benyoucef (2012); Liao and Kao (2011); Guo et al. (2009); Lee et al. (2015); Wu et al. (2016); Galankashi et al. (2016); Rao et al. (2017); Kaur et al. (2016)
12	Desire to do business	Dickson (1966); Bottani and Rizzi (2008); Wu et al. (2016)
13	Management and organization	Rouyendegh and Saputro (2014)
14	Operating controls	Dickson (1966); Guo et al. (2009)
15	Repair services • (After sales service) • (Service adaptability) • (Services)	Deng et al. (2014); Qian (2014); Kannan et al. (2014); Shen and Yu (2013); Junior et al. (2013); Mukherjee and Kar (2013); Chen and Wu (2013); Dey et al. (2012); Peng (2012); Amindoust et al. (2012); Zouggari and Benyoucef (2012); Liao and Kao (2010); Wang (2010); Ebrahim et al. (2009); Montazer et al. (2009); Guo et al. (2009); Ayhan and Kilic (2015); Akman (2015); Wu et al. (2016); Galankashi et al. (2016); Luthra et al. (2017); Ahmadi et al. (2017); Chaharsooghi and Ashrafi (2014); Fallahpour et al. (2017); Mehregan et al. (2014); Kaur et al. (2016); Govindan et al. (2013)
16	Attitude	Dickson (1966); Guo et al. (2009)
17	Flexibility	Junior et al. (2013); Ruiz-Torres et al. (2013); Erdem and Göçen (2012); Ahmadi et al. (2017); Zolghadri et al. (2011); Sanayei et al. (2010); Sanayei et al. (2010); Bai and Sarkis (2010); Montazer et al. (2009); Sanayei et al. (2008); (Kaur et al., 2016); Mehregan et al. (2014); Galankashi et al. (2016); Chaharsooghi and Ashrafi (2014); Fallahpour et al. (2017); Mani et al. (2016); Kaur et al. (2016); Zimmer et al. (2016)
18	Geographical location • (Distance to manufacturer)	Rouyendegh and Saputro (2014); Ghorbani et al. (2012); Erdem and Göçen (2012); Zouggari and Benyoucef (2012); Hsu et al. (2012); Kilincci and Onal (2011); Dalalah et al.

(Continued)

TABLE 3.5 (Continued)
Sustainability Criteria in Economic Dimension for Supplier Evaluation

S.No.	Economic criteria	Authors
		(2011); Aksoy and Öztürk (2011); Hassanzadeh and Razmi (2009); Dogan and Aydin (2011); Aydın Keskin et al. (2010); Guo et al. (2009); Lee et al. (2015); Galankashi et al. (2016); Kaur et al. (2016); Kannan (2018); Ghadimi et al. (2017)
19	Amount of past business	Zouggari and Benyoucef (2012)
20	Training Aid	Dickson (1966); Guo et al. (2009); Lee et al. (2015); Wu et al. (2016)
21	Reciprocal arrangement	Dickson (1966); Guo et al. (2009)
22	Reliability	Ruiz-Torres et al. (2013); Dey et al. (2012); Zolghadri et al. (2011); Hassanzadeh and Razmi (2009); Sanayei et al. (2008); Huang and Keskar (2007); Wu et al. (2016)
23	Impression	Dickson (1966)
24	Packaging ability	Erdem and Göçen (2012); Dalalah et al. (2011); Aydın Keskin et al. (2010); Guo et al. (2009)
25	Professionalism	Kilincci and Onal (2011)
26	Process Improvement	Dogan and Aydin (2011); Hsu and Hu (2009)
27	Product Development	Erdem and Göçen (2012); Guo et al. (2009)
28	JIT (Just in Time)	Dey et al. (2012); Aksoy and Öztürk (2011); Guo et al. (2009)
29	Commitment	Thiruchelvam and Tookey (2011)
30	Integrity	Thiruchelvam and Tookey (2011)
31	Long Term Relationship • (Joint venture) • (Collaboration) • (Partnership)	Dey et al. (2012); Devi et al. (2012); Liao and Kao (2011); Bai and Sarkis (2010); Ebrahim et al. (2009); Emre et al. (2009); Shen and Yu (2009); Guo et al. (2009); Lee et al. (2015); Kannan (2018); Zimmer et al. (2016)
32	Political Situation	Zouggari and Benyoucef (2012); Hsu et al. (2012); Montazer et al. (2009)

b. **Fuzzy rules (KM based)**

The team of DM then frames the rules for each sustainability criterion as per current industry practices. The rules framed by DMs for quality criteria and delivery time has been shown in Table 3.11. Similarly, all the criteria under the must be category are considered individually and rules are framed for FIS in each dimension. A sample of the assessment values is shown in Table 3.12.

c. **Defuzzification**

The Centre of Area (COM) method (Equation 3.3) is applied for defuzzification, and crisp values are obtained, which represent the EOPI, ENPI, and SOPI

TABLE 3.6

Sustainability Criteria in the Social Dimension for Supplier Evaluation

S.No.	Social Criteria	Authors
1	Social responsibility	Tavana et al. (2016); Kaur et al. (2016); Lee et al. (2009)
2	Ethical issues and legal complaint	Tavana et al. (2016)
3	Human rights • (Interests and rights of the employee)	Amindoust et al. (2012); Fallahpour et al. (2017); Luthra et al.(2017); Mani et al. (2016); Baskaran et al. (2012)
4	Rights of stakeholders(Involvement of stakeholders)	Amindoust et al. (2012); Zimmer et al. (2016); Luthra et al.(2016)
5	Health and safety of employees • (Health and safety) • (Commitment to health and safety of employees)	Amindoust et al. (2012); Rao et al. (2017); Govindan et al. (2013); Fallahpour et al. (2017); Luthra et al. (2017); Orji and Wei (2015); Mehregan et al. (2014); Bai and Sarkis (2010); Ahmadi et al. (2017); Azadnia et al. (2013); Mani et al. (2016); Winter and Lasch (2016); Husgafvel et al. (2015); Zimmer et al. (2016); Tavana et al. (2016)
6	Information disclosure	Amindoust et al. (2012); Luthra et al. (2017); Orji and Wei (2015)
7	Respect for the policy	Amindoust et al. (2012); Orji and Wei (2015); Mehregan et al. (2014)
8	Employment practices	Rao et al. (2017); Govindan et al. (2013); Bai and Sarkis (2010); Ghadimi and Heavey (2014)
9	Local community influence	Rao et al. (2017); Govindan et al. (2013); Bai and Sarkis (2010); Ahmadi et al. (2017)
10	Stakeholders influence (stakeholder relations)	Rao et al. (2017); Govindan et al. (2013); Mehregan et al. (2014); Zimmer et al. (2016); Ahmadi et al. (2017)
11	Enterprise Resource Planning (ERP) system	Kaur et al. (2016)
12	Human resource capability	Mehregan et al. (2014)
13	Child and bonded labour	Mani et al. (2016); Winter and Lasch (2016)
14	Wages	Mani et al. (2016)
15	Society	Mani et al. (2016)
16	Regulatory responsibility	Mani et al. (2016)
17	Annual number of accidents	Zimmer et al. (2016)

of each supplier (Table 3.13). Based on the values of EOPI, it elicits that for the economic dimension, supplier 3 has the highest performance. ENPI is highest for supplier 7 and the SOPI of supplier 1 is highest. The Values of EOPI, ENPI, and SOPI are added to calculate the Total Performance Index (TPI) for every supplier (Table 3.14).

TABLE 3.7
Sustainability Criteria in Environmental Dimension for Supplier Evaluation

S.No	Environmental criteria	Authors
1	Green Image • (Green corporate social image)	Ahmadi et al. (2017); Shen et al. (2013); Lee et al. (2009)
2	Eco-design • (Green design)	Ahmadi et al. (2017); Chaharsooghi and Ashrafi (2014); Amindoust et al. (2012); Luthra et al. (2017); Orji and Wei (2015); Kannan (2018); Govindan et al. (2013); Azadi et al. (2015); Zimmer et al. (2016); Shen et al. (2013); Shaik and Abdul-Kader (2011); Teixeira et al. (2016); Kannan et al. (2013); Uygun and Dede (2016)
3	Environmental management system (EMS) • (environmental responsibility) • (environment adaptability) • (environmental competencies) • (environmental code of conduct) • (environmental control certification)	Ahmadi et al. (2017); Chaharsooghi and Ashrafi (2014); Azadnia et al. (2015); Fallahpour et al. (2017); Mehregan et al. (2014); Amindoust et al. (2012); Bai and Sarkis (2010); Luthra et al. (2017); Rao et al. (2017); Kaur et al. (2016); Tavana et al. (2016); Orji and Wei (2015); Kannan (2018); Govindan et al. (2013); Ghadimi et al. (2017); Azadi et al. (2015); Azadnia et al. (2013); Azadnia et al. (2015); Lee et al. (2009); Teixeira et al. (2016); Kannan et al. (2013)
4	End-of-pipe	Ahmadi et al. (2017)
5	Pollution Control • (Pollution production) • (Pollution Prevention)	Chaharsooghi and Ashrafi (2014); Azadnia et al. (2015); Amindoust et al. (2012); Bai and Sarkis (2010); Tavana et al. (2016); Spangenberg (2004); Govindan et al. (2013); Azadi et al. (2015); Azadnia et al. (2013); Azadnia et al. (2015); Shen et al. (2013); Kannan et al. (2013)
6	Resource consumption,	Chaharsooghi and Ashrafi (2014); Govindan et al. (2013); Zimmer et al. (2016); Shen et al. (2013); Kannan et al. (2013)
7	Greenhouse emission • (Greenhouse effect)	Azadnia et al.(2015); Neumüller et al. (2016); Ghadimi et al. (2017); Azadnia et al. (2013); Azadnia et al. (2015); Zimmer et al. (2016)
8	Green Packing and Labelling	Fallahpour et al. (2017); Amindoust et al. (2012); Luthra et al. (2017); Tavana et al. (2016); Ghadimi et al. (2017); Azadi et al. (2015); Gurel et al. (2015); Lee et al. (2009); Teixeira et al. (2016)
9	Green warehousing	Fallahpour et al. (2017)
10	Green technology • (Green R & D) • (Green manufacturing) • (Green process planning)	Fallahpour et al. (2017); Amindoust et al. (2012); Luthra et al. (2017); Kannan (2018); Azadi et al. (2015); Shen et al. (2013); Lee et al. (2009); Shaik and Abdul-Kader (2011)
11	Green transportation • (Green logistics)	Fallahpour et al. (2017); Uygun and Dede (2016),

TABLE 3.7 (Continued)
Sustainability Criteria in Environmental Dimension for Supplier Evaluation

S.No	Environmental criteria	Authors
12	Energy management system • (Energy consumption)	Mehregan et al. (2014), Neumüller et al. (2016), Azadi et al. (2015), Zimmer et al. (2016), Gurel et al. (2015), Lee et al. (2009)
13	hazardous material management • (Harmful chemical) • (Waste Material management) • (Waste Production) • (Solid waste management)	Amindoust et al. (2012), Luthra et al. (2017), Neumüller et al. (2016), Rao et al. (2017), Tavana et al. (2016), Kannan (2018), Ghadimi et al. (2017), Azadi et al. (2015), Gurel et al. (2015), Lee et al. (2009), Shaik and Abdul-Kader (2011)
14	Recycling/ Reuse / Remanufacture	Amindoust et al. (2012), Rao et al. (2017), Kannan (2018), Ghadimi et al. (2017), Zimmer et al. (2016), Gurel et al. (2015), Lee et al. (2009)
15	Noise	Rao et al. (2017)
16	Carbon footprint tax	Kaur et al. (2016), Rao et al. (2017)
17	Environmental training • (Green training)	Shen et al. (2013), Teixeira et al. (2016)

TABLE 3.8
Response of FKQ for Quality Criteria

Fuzzy Kano model questionnaire (multiple responses)					
	Like	Must Be	Neutral	Live with	Dislike
How you feel if quality is considered for the SSS process	40%	60%			
How you feel if quality not considered as SSS process			20%	20%	60%

Step 4: FIS model considering Attractive sustainability criteria for final selection

In this step, the final selection of the suppliers is done based on the evaluation against the sustainability criteria under the Attractive fuzzy Kano cluster. In the final selection stage, four suppliers are considered from the pre-selection stage. In this step, three distinct FIS engines are developed with a modified structure to have the least number of rules to be framed by DM. Three discrete FIS for TBL dimensions are made for the supplier's performance. For each FIS, quantified evaluation levels of the Attractive criteria act as the inputs for the

TABLE 3.9

Kano Clusters for Sustainability Criteria

Sustainability criteria in the economic dimension	Fuzzy Kano clusters	
	Must be	Attractive
	• Quality	• Warranties and claim policies
	• Delivery	• Amount of past business
	• Production facilities	• Training Aid
	• Net Price	• Reciprocal arrangement
	• Technical capability	• Packaging ability
	• Repair services	• Product development
		• JIT
		• Long term relationship
		• Political situation
Sustainability criteria in environmental dimension	• Eco-design	• End of pipe
	• Environmental management system	• Green warehousing
	• Pollution control	• Green transportation
	• Green technology	• Carbon footprint tax
	• Energy management system	• Environmental training
	• Hazardous material management	
Sustainability criteria in the social dimension	• Social responsibility	• Human resource capability
	• Human rights	• Wages
	• Rights of stakeholders.	• Society
	• Health and safety of employees	
	• Employment practices	
	• Stakeholder influence	

FIS. The outputs of three FIS are the PI of the suppliers in three sustainability dimensions for the final selection stage.

a. **Membership function and operational operators**

This stage FISalso utilizes a triangular membership function for all three FIS engines. Similar to the pre-selection stage here also Performance Index for PIED, PISD and PIEND are evaluated using three linguistic variables.

b. *Fuzzy rules (KM based) and Modified Architecture of FIS engine*

In this step, using DM's experience and knowledge, fuzzy rules are recognized in all the three TBL dimensions similar to the pre-selection stage. Further, all

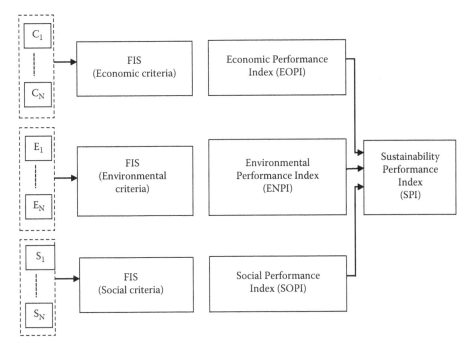

FIGURE 3.4 FIS model for pre-selection of supplier considering must be criteria

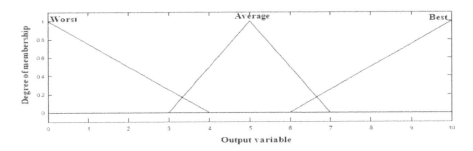

FIGURE 3.5 Membership function for suppliers performance target range

potential alternatives are evaluated using these rules through FIS models (Table 3.15).

c. *Defuzzification*

The output of all distinct FIS is defuzzified using the COA method as done in the pre-selection stage (Table 3.15). It can be seen that S2 out marks all other suppliers in the economic dimension and the S1 has the lowest performance. The weights of the three dimensions as finalized by DM's are 0.5 (economic), 0.3 (environmental),

TABLE 3.10
Evaluation Levels for Quality Criteria

Evaluation level Description

Evaluation level	Description
Best	There is no defective item in the supplied order lot or is less than 1%
Average	The percentage of defective items/material supplied is more than 1% but less than 3%
Worst	The percentage of defective items/material supplied is more than 3%

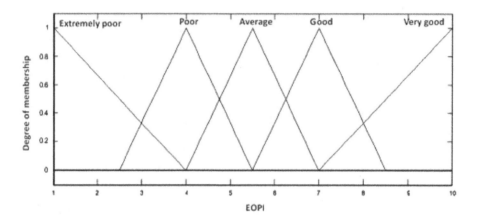

FIGURE 3.6 Performance index target range

TABLE 3.11
FIS Rules for Quality and Delivery Time

Rule No. Rules

Rule No.	Rules
R1	When % of Quality supplied is best AND Delivery time is best then the output is best
R2	When %of Quality supplied is best AND Delivery time is Average then the output is best
R3	When %of Quality supplied is best AND Delivery time is Worst then the output is Average
R4	When %of Quality supplied is Average AND Delivery time is best, then the output is best
R5	When %of Quality supplied is Average AND Delivery time is best then the output is Average
R6	When %of Quality supplied is Average AND Delivery time is Worst then the output is Worst
R7	When %of Quality supplied is Worst AND Delivery time is best then the output is Average
R8	When %of Quality supplied is Worst AND Delivery time is best then the output is Worst
R9	When %of Quality supplied is Worst AND Delivery time is best then the output is Worst

TABLE 3.12

Evaluation Values for Suppliers

Sustainability criteria in the economic dimension

	S1	S2	S3	S4	S5	S6	S7
C1	0.5	2	3	2.5	2	0.8	2
C2	1	2	3	0	2	4	1
C3	-9	-2	-10	-10	-3	-7	4
C4	9	5	8	7	3	8	8
C5	0.96	0.9	0.99	1	0.98	0.92	0.94
C6	0.96	0.94	0.95	0.9	0.97	1	0.92
Sustainability criteria in the social dimension							
S1& S2	9	7	5	7	9	8	9
S3	3	0	2	1.5	2	3	3
S5	9	6	5	6	8	7	8
S6	9	6	5	6	8	7	3.5
Sustainability criteria in environmental dimension							
E1 & E2	8	9	5	6	4	9.5	8
E3 & E4	8	9	6	7	4	9.5	8
E5	7	8	5	5.5	5	9.5	8
E6	7	8	6	6.5	5	9.5	8

TABLE 3.13

Performance Index of Suppliers in Three TBL Dimensions

	EOPI	ENPI	SOPI	TPI	SPI	Ranks
S1	8.568	8.908	8.929	26.405	8.742	1
S2	1.449	8.926	3.999	14.374	4.202	7
S3	8.570	7.001	5.500	21.071	7.485	4
S4	5.000	8.873	7.001	20.874	6.562	5
S5	8.563	5.500	8.918	22.981	7.715	3
S6	8.537	8.918	8.892	26.347	8.722	2
S7	1.509	8.942	8.730	19.181	5.183	6

and 0.2 (social). Further, using Equation 3.4, the performance of every potential supplier is calculated, and ranking is awarded to suppliers. The final ranks of the suppliers are S2 > S4 > S3 > S1.

TABLE 3.14

Suppliers Assessment Values for Attractive Cluster Criteria by Decision-makers

Sustainability criteria in the Economic dimension

Criteria	S1	S2	S3	S4
Warranties and claim policies	75	90	65	80
Amount of past business	20	24	10	15
Long term relationship	5	10	8	6
Product development	8	10	5	9
JIT	1.02	1.01	1.04	1.0
Reciprocal arrangement	5	13	11	12
Packaging ability	10	70	60	90
Sustainability criteria in the Environmental dimension				
Green warehousing	1	9	7	6
Carbon footprint tax	2	8	5	5
Green transportation	10	70	10	10
Sustainability criteria in the Social dimension				
Wages	85	85	75	95
Human resource capability	12	12	5	14

3.5   MANAGERIAL IMPLICATIONS

In the proposed work, a three-stage sustainable supplier selection framework based on the FKM and Fuzzy inference system has been proposed. The results of the proposed integrated methodology are of importance for the supply chain managers for making effective decisions for supplier selection in the I40 horizon. The proposed work gives a better insight into the sustainability performance of supplier's for both pre-selection and final selection. The proposed work can be applied by the supply chain managers of other industries by selecting the appropriate sustainability criteria of the industry under consideration.

3.6   CONCLUSIONS AND FUTURE WORK

Industry 4.0 has empowered industries with a greater capacity of producing quality products at a very high production rate. Supply chain managers now have great responsibility to achieve the sustainability aspects of the entire supply chain of the industry. Selecting suppliers on sustainability dimensions can help industries establish sustainability objectives in the supply chain. The three-phase SSS framework aids the DMs in selecting the most proficient sustainable supplier among the alternatives. Two separate FIS models have been developed for pre-selection and final selection of the suppliers. The pre-selection stage has been accomplished by considering the Must-be criterion for supplier assessment and for the final selection stage, Attractive sustainability criteria in each TBL dimension have been considered.

TABLE 3.15

Suppliers Performance Index Values in all Sustainability Dimensions Through FIS

Supplier	PIED				PIEND)				PISD			SPI	Overall Ranks
	Calculated values	Normalized values	Ranks	Calculated values	Normalized values	Ranks	Calculated values	Normalized values	Ranks				
1	3.99	0.15	4	2.08	0.10	4	1.56	0.081	3	0.125	4		
2	8.91	0.35	1	8.90	0.43	1	7.72	0.401	2	0.386	1		
3	7	0.27	2	5.5	0.26	2	1.48	0.077	4	0.233	3		
4	5.5	0.21	3	3.99	0.19	3	8.47	0.44	1	0.254	2		

The framework facilitates the DM to decide the weightage of each TBL dimension as per the need of the industry. The proposed framework has been applied to a large-scale industry. As future work, researchers can apply the framework to small- and medium-scale industries also. Further researchers can apply the MCDM techniques and can compare the results of the FIS model for analysis.

REFERENCES

Ahmadi, H.B., Hamid, S., Petrudi, H., and Wang, X. (2017) Integrating sustainability into supplier selection with analytical hierarchy process and improved grey relational analysis: a case of telecom industry. *The International Journal of Advanced Manufacturing Technology*, 90(9–12), pp. 2413–2427. doi: 10.1007/s00170-016-9518-z

Akman, G. (2015) Evaluating suppliers to include green supplier development programs via fuzzy c-means and VIKOR methods. *Computers & Industrial Engineering*, 86, pp. 69–82. doi: 10.1016/j.cie.2014.10.013

Aksoy, A., and Öztürk, N. (2011) Supplier selection and performance evaluation in just-in-time production environments. *Expert Systems with Applications*, 38, pp. 6351–6359. doi: 10.1016/j.eswa.2010.11.104.

Aksoy, A., Sucky, E., and Öztürk, N. (2014) Dynamic strategic supplier selection system with fuzzy logic. *Procedia - Social and Behavioral Sciences*, 109, pp. 1059–1063. doi: 10.1016/j.sbspro.2013.12.588

Amindoust, A., Ahmed, S., Saghafinia, A., and Bahreininejad, A. (2012) Sustainable supplier selection: a ranking model based on fuzzy inference system. *Applied Soft Computing*, 12(6), pp. 1668–1677. doi: 10.1016/j.asoc.2012.01.023

Araz, C., and Ozkarahan, I. (2007) Supplier evaluation and management system for strategic sourcing based on a new multicriteria sorting procedure. *International Journal of Production Economics*, 106(2), pp. 585–606. doi: 10.1016/j.ijpe.2006.08.008

Arikan, F. (2013) A fuzzy solution approach for multi objective supplier selection. *Expert Systems with Applications*, 40(3), pp. 947–952. doi: 10.1016/j.eswa.2012.05.051

Aydın Keskin, G., İlhan, S., and Özkan, C. (2010) The fuzzy ART algorithm: a categorization method for supplier evaluation and selection. *Expert Systems with Applications*, 37(2), pp. 1235–1240. doi: 10.1016/j.eswa.2009.06.004

Ayhan, M.B., and Kilic, H.S. (2015) A two stage approach for supplier selection problem in multi-item/multi-supplier environment with quantity discounts. *Computers and Industrial Engineering*, 85, pp. 1–12. doi: 10.1016/j.cie.2015.02.026

Azadi, M., Jafarian, M., Saen, R.F., and Mirhedayatian, S.M. (2015) A new fuzzy DEA model for evaluation of efficiency and effectiveness of suppliers in sustainable supply chain management context. *Computers and Operations Research*, 54(July 2015), pp. 274–285. doi 10.1016/j.cor.2014.03.002

Azadnia, A.H., Ghadimi, P., Saman, M.Z.M., Wong, K.Y., and Heavey, C. (2013) An integrated approach for sustainable supplier selection using fuzzy logic and fuzzy AHP. *Applied Mechanics and Materials*, 315(April), pp. 206–210. doi: 10.4028/www.scientific.net/AMM.315.206

Azadnia, Amir Hossein, Saman, M.Z.M., and Wong, K.Y. (2015) Sustainable supplier selection and order lot-sizing: an integrated multi-objective decision-making process. *International Journal of Production Research*, 53(2), pp. 383–408. doi: 10.1080/00207543.2014.935827

Azadnia, Amir Hossein, Saman, M.Z.M., Wong, K.Y., Ghadimi, P., and Zakuan, N. (2012) Sustainable supplier selection based on self-organizing map neural network and multi criteria decision making approaches. *Procedia - Social and Behavioral Sciences*, 65, pp. 879–884. doi: 10.1016/j.sbspro.2012.11.214

Bai, C., and Sarkis, J. (2010) Integrating sustainability into supplier selection with grey system and rough set methodologies. *International Journal of Production Economics*, 124(1), pp. 252–264. doi: 10.1016/j.ijpe.2009.11.023

Baskaran, V., Nachiappan, S., and Rahman, S. (2012) Indian textile suppliers sustainability evaluation using the grey approach. *International Journal of Production Economics*, 135(2), pp. 647–658. doi: 10.1016/j.ijpe.2011.06.012

Beikkhakhian, Y., Javanmardi, M., Karbasian, M., and Khayambashi, B. (2015) The application of ISM model in evaluating agile suppliers selection criteria and ranking suppliers using fuzzy TOPSIS-AHP methods. *Expert Systems with Applications*, 42(15–16), pp. 6224–6236. doi: 10.1016/j.eswa.2015.02.035

Bonilla, S.H., Silva, H.R.O., Terra da Silva, M., Franco Gonçalves, R., and Sacomano, J.B. (2018) Industry 4.0 and sustainability implications: A scenario-based analysis of the impacts and challenges. *Sustainability*, 10(10), Art. no. 3740.

Bottani, E., and Rizzi, A. (2008) An adapted multi-criteria approach to suppliers and products selection—An application oriented to lead-time reduction. *International Journal of Production Economics*, 111(2), pp. 763–781. doi: 10.1016/j.ijpe.2007.03.012

Braccini, A., and Margherita, E. (2018) Exploring organizational sustainability of Industry 4.0 under the triple bottom line: the case of a manufacturing company. *Sustainability*, 11(1), Art. no. 36. doi: 10.3390/su11010036

Chaharsooghi, S.K., and Ashrafi, M. (2014) Sustainable supplier performance evaluation and selection with neofuzzy TOPSIS method. *International Scholarly Research Notices*, 2014, pp. 1–10.

Che, Z.H., and Wang, H.S. (2010) A hybrid approach for supplier cluster analysis. *Computers and Mathematics with Applications*, 59(2), pp. 745–763. doi: 10.1016/j.camwa.2009.10.018

Chen, L.-H., and Ko, W.-C. (2008) A fuzzy nonlinear model for quality function deployment considering Kano's concept. *Mathematical and Computer Modelling*, 48(3–4), pp. 581–593. doi: 10.1016/j.mcm.2007.06.029

Chen, P.-S., and Wu, M.-T. (2013) A modified failure mode and effects analysis method for supplier selection problems in the supply chain risk environment: A case study. *Computers & Industrial Engineering*, 66(4), pp. 634–642. doi: 10.1016/j.cie.2013.09.018

Chyu, C., and Fang, Y. (2014) A hybrid fuzzy analytic network process approach to the new product development selection problem. *Mathematical Problems in Engineering*, 2014, pp. 1–13.

Dalalah, D., Hayajneh, M., and Batieha, F. (2011) Expert systems with applications a fuzzy multi-criteria decision making model for supplier selection. *Expert Systems With Applications*, 38(7), pp. 8384–8391. doi: 10.1016/j.eswa.2011.01.031

Demir, K.A., Döven, G., and Sezen, B. (2019) Industry 5.0 and human-robot co-working. *Procedia Computer Science*, 158, pp. 688–695. doi: 10.1016/j.procs.2019.09.104

Deng, S., Aydin, R., Kwong, C.K., and Huang, Y. (2014) Integrated product line design and supplier selection: a multi-objective optimization paradigm. *Computers & Industrial Engineering*, 70, pp. 150–158. doi: 10.1016/j.cie.2014.01.011

Deng, X., Hu, Y., Deng, Y., and Mahadevan, S. (2014) Supplier selection using AHP methodology extended by D numbers. *Expert Systems with Applications*, 41(1), pp. 156–167. doi: 10.1016/j.eswa.2013.07.018

Dey, S., Kumar, A., Ray, A., and Pradhan, B.B. (2012) Supplier selection: integrated theory using DEMATEL and quality function deployment methodology. *Procedia Engineering*, 38, pp. 3560–3565. doi: 10.1016/j.proeng.2012.06.411

Dickson, G.W. (1966) An analysis of vendor selection and the buying process. *Journal of Purchasing*, 2(1), pp. 5–17.

Dogan, I., and Aydin, N. (2011) Combining Bayesian networks and total cost of ownership method for supplier selection analysis. *Computers & Industrial Engineering*, 61(4), pp. 1072–1085. doi: 10.1016/j.cie.2011.06.021

Dweiri, F., Kumar, S., Khan, S.A., and Jain, V. (2016) Designing an integrated AHP based decision support system for supplier selection in automotive industry. *Expert Systems with Applications*, 62, pp. 273–283. doi: 10.1016/j.eswa.2016.06.030

Ebrahim, R.M., Razmi, J., and Haleh, H. (2009) Advances in Engineering Software Scatter search algorithm for supplier selection and order lot sizing under multiple price discount environment. *Advances in Engineering Software*, 40(9), pp. 766–776. doi: 10.1016/j.advengsoft.2009.02.003

Ekici, A. (2013) An improved model for supplier selection under capacity constraint and multiple criteria. *International Journal of Production Economics*, 141(2), pp. 574–581. doi: 10.1016/j.ijpe.2012.09.013

Emre, F., Genç, S., Kurt, M., and Akay, D. (2009) A multi-criteria intuitionistic fuzzy group decision making for supplier selection with TOPSIS method. *Expert Systems With Applications*, 36(8), pp. 11363–11368. doi: 10.1016/j.eswa.2009.03.039

Erdem, A.S., and Göçen, E. (2012) Development of a decision support system for supplier evaluation and order allocation. *Expert Systems with Applications*, 39(5), pp. 4927–4937. doi: 10.1016/j.eswa.2011.10.024

Fallahpour, A., Olugu, E.U., Musa, S.N., Wong, K.Y., and Noori, S. (2017) A decision support model for sustainable supplier selection in sustainable supply chain management. *Computers & Industrial Engineering*, 105, pp. 391–410. doi: 10.1016/j.cie.2017.01.005

Florez-Lopez, R., and Ramon-Jeronimo, J.M. (2012) Managing logistics customer service under uncertainty: an integrative fuzzy Kano framework. *Information Sciences*, 202, pp. 41–57. http://dx.doi.org/10.1016/j.ins.2012.03.004

Gabriel, M., and Pessl, E. (2016) Industry 4.0 and sustainability impacts: critical discussion of sustainability aspects with a special focus on future of work and ecological consequences. *Annals of the Faculty of Engineering Hunedoara*, 14(2), pp. 131–136.

Galankashi, M.R., Helmi, S.A., and Hashemzahi, P. (2016) Supplier selection in automobile industry: a mixed balanced scorecard-fuzzy AHP approa ch. *Alexandria Engineering Journal*, 55(1), pp. 93–100. doi: 10.1016/j.aej.2016.01.005

García, N., Puente, J., Fernández, I., and Priore, P. (2013) Supplier selection model for commodities procurement. Optimised assessment using a fuzzy decision support system. *Applied Soft Computing*, 13(4), pp. 1939–1951. doi: 10.1016/j.asoc.2012.12.008

Ghadimi, P., Dargi, A., and Heavey, C. (2017) Sustainable supplier performance scoring using audition check-list based fuzzy inference system: a case application in automotive spare part industry. *Computers & Industrial Engineering*, 105, pp. 12–27. doi: 10.1016/j.cie.2017.01.002

Ghadimi, P., and Heavey, C. (2014) Sustainable supplier selection in medical device industry: Toward sustainable manufacturing. *Procedia CIRP*, 15, pp. 165–170. doi: 10.1016/j.procir.2014.06.096

Ghadimi, P., Wang, C., Lim, M.K., and Heavey, C. (2019) Intelligent sustainable supplier selection using multi-agent technology: theory and application for Industry 4.0 supply chains. *Computers and Industrial Engineering*, 127(October), pp. 588–600. doi: 10.1016/j.cie.2018.10.050

Ghorbani, M., Bahrami, M., and Arabzad, S.M. (2012) An integrated model for supplier selection and order allocation; using Shannon entropy, SWOT and linear Programming. *Procedia - Social and Behavioral Sciences*, 41(1), pp. 521–527. doi: 10.1016/j.sbspro.2012.04.064

Govindan, K., Khodaverdi, R., and Jafarian, A. (2013) A fuzzy multi criteria approach for measuring sustainability performance of a supplier based on triple bottom line approach. *Journal of Cleaner Production*, 47, pp. 345–354.

Guneri, A.F., Yucel, A., and Ayyildiz, G. (2009) An integrated fuzzy-lp approach for a supplier selection problem in supply chain management. *Expert Systems With Applications*, 36(5), pp. 9223–9228. doi: 10.1016/j.eswa.2008.12.021

Guo, X., Yuan, Z., and Tian, B. (2009) SupAttractive quality and must-be qualityplier selection based on hierarchical potential support vector machine. *Expert Systems With Applications*, 36(3), pp. 6978–6985. 10.1016/j.eswa.2008.08.074

Guo, X., Zhu, Z., and Shi, J. (2014) Integration of semi-fuzzy SVDD and CC-Rule method for supplier selection. *Expert Systems with Applications*, 41(4), pp. 2083–2097. doi: 10.1016/j.eswa.2013.09.008

Gurel, O., Acar, A.Z., Onden, I., and Gumus, I. (2015) Determinants of the green supplier selection. *Procedia - Social and Behavioral Sciences*, 181, pp. 131–139. doi: 10.1016/j.sbspro.2015.04.874

Hannan, A., Ahmad, J., and Basit, A. (2015) Value based requirements classification of software product using fuzzy Kano model. *New Horrizon*, 83, pp. 48–56.

Hassanzadeh, S., and Razmi, J. (2009) An integrated fuzzy model for supplier management: a case study of ISP selection and evaluation. *Expert Systems with Applications*, 36(4), pp. 8639–8648. doi: 10.1016/j.eswa.2008.10.012

Hermann, M., Pentek, T., and Otto, B. (2016) Design principles for industrie 4.0 scenarios. In *Proceedings of the Annual Hawaii International Conference on System Sciences,* pp. 3928–3937. doi: 10.1109/HICSS.2016.488

Herrmann, C., Schmidt, C., Kurle, D., Blume, S., and Thiede, S. (2014) Sustainability in manufacturing and factories of the future. *International Journal of Precision Engineering and Manufacturing-Green Technology*, 1(4), pp. 283–292. doi: 10.1007/s40684-014-0034-z

Horváth, D., and Szabó, R.Z. (2019) Driving forces and barriers of Industry 4.0: do multinational and small and medium-sized companies have equal opportunities? *Technological Forecasting and Social Change*, 146, pp. 119–132. https://doi.org/10.1016/j.techfore.2019.05.021

Hsu, C.H., Wang, F.K., and Tzeng, G.H. (2012) The best vendor selection for conducting the recycled material based on a hybrid MCDM model combining DANP with VIKOR. *Resources, Conservation and Recycling*, 66(2012), pp. 95–111. doi: 10.1016/j.resconrec.2012.02.009

Hsu, C., and Hu, A.H. (2009) Applying hazardous substance management to supplier selection using analytic network process. *Journal of Cleaner Production*, 17(2), pp. 255–264. doi: 10.1016/j.jclepro.2008.05.004

Huang, L.F., and Wu, B. (1992) Analysis of fuzzy statistics and its applications in surveys. Taiwan: Department of Statistics, National Cheng Chi University.

Huang, S.H., and Keskar, H. (2007) Comprehensive and configurable metrics for supplier selection. *International Journal of Production Economics*, 105(2), pp. 510–523. doi: 10.1016/j.ijpe.2006.04.020

Husgafvel, R., Pajunen, N., Virtanen, K., Paavola, I.-L., Päällysaho, M., Inkinen, V., … Ekroos, A. (2015) Social sustainability performance indicators – experiences from process industry. *International Journal of Sustainable Engineering*, 8(1), pp. 14–25. doi: 10.1080/19397038.2014.898711

Jadidi, O., Zolfaghari, S., and Cavalieri, S. (2014) A new normalized goal programming model for multi-objective problems: A case of supplier selection and order allocation. *International Journal of Production Economics*, 148, pp. 158–165. doi: 10.1016/j.ijpe.2013.10.005

Jain, Naveen, and Singh, A.R. (2019) Sustainable supplier selection criteria classification for Indian iron and steel industry: a fuzzy modified Kano model approach. *International Journal of Sustainable Engineering*, 13, pp. 1–16. doi: 10.1080/19397038.2019.1566413.

Jain, Naveen, and Singh, A.R. (2020) Sustainable supplier selection under must-be criteria through Fuzzy inference system. *Journal of Cleaner Production*, 248, Art. no. 119275. doi: 10.1016/j.jclepro.2019.119275

Jain, Naveen, Singh, A.R., and Choudhary, A.K. (2016) Integrated methodology for supplier selection in supply chain management. In *2016 IEEE International Conference on Industrial Engineering and Engineering Management (IEEM)*, pp. 807–811. doi: 10.11 09/IEEM.2016.7797988

Junior, F.R.L., Osiro, L., and Carpinetti, L.C.R. (2013) A fuzzy inference and categorization approach for supplier selection using compensatory and non-compensatory decision rules. *Applied Soft Computing Journal*, 13(10), pp. 4133–4147. doi: 10.1016/j.asoc.2013.06.020

Kamble, S.S., Gunasekaran, A., and Gawankar, S.A. (2018) Sustainable Industry 4.0 framework: a systematic literature review identifying the current trends and future perspectives. *Process Safety and Environmental Protection*, 117, pp. 408–425. doi: 10.101 6/j.psep.2018.05.009

Kannan, D. (2018) Role of multiple stakeholders and the critical success factor theory for the sustainable supplier selection process. *International Journal of Production Economics*, 195, pp. 391–418. doi: 10.1016/j.ijpe.2017.02.020

Kannan, D., Govindan, K., and Rajendran, S. (2014) Fuzzy axiomatic design approach based green supplier selection: a case study from Singapore. *Journal of Cleaner Production*, 96, pp. 194–208. http://dx.doi.org/10.1016/j.jclepro.2013.12.076

Kannan, D., Khodaverdi, R., Olfat, L., Jafarian, A., and Diabat, A. (2013) Integrated fuzzy multi criteria decision making method and multi-objective programming approach for supplier selection and order allocation in a green supply chain. *Journal of Cleaner Production*, 47, pp. 355–367. doi: 10.1016/j.jclepro.2013.02.010

Kano, N., Seraku, N., Takahashi, F., and Tsuji, S. (1984) Attractive quality and must-be quality. *The Journal of Japanese Society for Quality Control*, 14, pp. 39–48.

Kar, A.K. (2014) Revisiting the supplier selection problem: an integrated approach for group decision support. *Expert Systems with Applications*, 41(6), pp. 2762–2771. doi: 10.101 6/j.eswa.2013.10.009

Karsak, E.E., and Dursun, M. (2015) An integrated fuzzy MCDM approach for supplier evaluation and selection. *Computers and Industrial Engineering*, 82, pp. 82–93. doi: 10.1016/j.cie.2015.01.019

Kaur, H., Singh, S.P., and Glardon, R. (2016) An integer linear program for integrated supplier selection: a sustainable flexible framework. *Global Journal of Flexible Systems Management*, 17(2), pp. 113–134. doi: 10.1007/s40171-015-0105-1

Khaleie, S., Fasanghari, M., and Tavassoli, E. (2012) Supplier selection using a novel intuitionist fuzzy clustering approach. *Applied Soft Computing Journal*, 12(6), pp. 1741–1754. doi: 10.1016/j.asoc.2012.01.017

Kiel, D., Müller, J.M., Arnold, C., and Voigt, K.-I. (2017) Sustainable industrial value creation: benefits and challenges of Industry 4.0. *International Journal of Innovation Management*, 21(8), Art. no. 1740015. doi: 10.1142/S1363919617400151

Kilincci, O., and Onal, S.A. (2011) Expert systems with applications fuzzy AHP approach for supplier selection in a washing machine company. *Expert Systems With Applications*, 38(8), pp. 9656–9664. doi: 10.1016/j.eswa.2011.01.159

Kumar, D., Singh, J., and Singh, O.P. (2013) A fuzzy logic based decision support system for evaluation of suppliers in supply chain management practices. *Mathematical and Computer Modelling*, 58(11–12), pp. 1679–1695. doi: 10.1016/j.mcm.2013.07.003

Lam, K.C., Tao, R., and Lam, M.C.K. (2010) A material supplier selection model for property developers using fuzzy principal component analysis. *Automation in Construction*, 19(5), pp. 608–618. doi: 10.1016/j.autcon.2010.02.007

Lasi, H., Fettke, P., Kemper, H.G., Feld, T., and Hoffmann, M. (2014) Industry 4.0. *Business and Information Systems Engineering*, 6(4), pp. 239–242. doi: 10.1007/s12599-014-0334-4.

Lee, A.H.I., Kang, H., Hsu, C., and Hung, H. (2009) A green supplier selection model for high-tech industry. *Expert Systems with Applications*, 36(4), pp. 7917–7927. doi: 10.1016/j.eswa.2008.11.052

Lee, J., Cho, H., and Kim, Y.S. (2015) Assessing business impacts of agility criterion and order allocation strategy in multi-criteria supplier selection. *Expert Systems with Applications*, 42(3), pp. 1136–1148. 10.1016/j.eswa.2014.08.041

Lee, Y.-C., and Huang, S.-Y. (2009) A new fuzzy concept approach for Kano's model. *Expert Systems with Applications*, 36(3), pp. 4479–4484. doi: 10.1016/j.eswa.2008.05.034

Lee, Y.-C., Sheu, L.-C., and Tsou, Y.-G. (2008) Quality function deployment implementation based on fuzzy Kano model: an application in PLM system. *Computers & Industrial Engineering*, 55(1), pp. 48–63. doi: 10.1016/j.cie.2007.11.014

Liao, C.-N., and Kao, H.-P. (2010) Supplier selection model using Taguchi loss function, analytical hierarchy process and multi-choice goal programming. *Computers & Industrial Engineering*, 58(4), pp. 571–577. doi: 10.1016/j.cie.2009.12.004

Liao, C., and Kao, H. (2011) An integrated fuzzy TOPSIS and MCGP approach to supplier selection in supply chain management. *Expert Systems with Applications*, 38(9), pp. 10803–10811. doi: 10.1016/j.eswa.2011.02.031

Lima-Junior, F.R., and Carpinetti, L.C.R. (2016) Combining SCOR® model and fuzzy TOPSIS for supplier evaluation and management. *International Journal of Production Economics*, 174, pp. 128–141. doi: 10.1016/j.ijpe.2016.01.023

Lima Junior, F.R., Osiro, L., and Carpinetti, L.C.R. (2014) A comparison between fuzzy AHP and fuzzy TOPSIS methods to supplier selection. *Applied Soft Computing*, 21, pp. 194–209. doi: 10.1016/j.asoc.2014.03.014

Luthra, S., Govindan, K., Kannan, D., Mangla, S.K., and Garg, C.P. (2017) An integrated framework for sustainable supplier selection and evaluation in supply chains. *Journal of Cleaner Production*, 140, pp. 1686–1698. doi: 10.1016/j.jclepro.2016.09.078

Mani, V., Gunasekaran, A., Papadopoulos, T., Hazen, B., and Dubey, R. (2016) Supply chain social sustainability for developing nations: evidence from India. *Resources, Conservation and Recycling*, 111(August), pp. 42–52. doi: 10.1016/j.resconrec.2016.04.003

Manski, C.F. (1990) The use of intentions data to predict behavior: a best-case analysis. *Journal of the American Statistical Association*, 85(412), pp. 934–940.

Mehregan, M.R., Hashemi, S.H., Karimi, A., and Merikhi, B. (2014) Analysis of interactions among sustainability supplier selection criteria using ISM and fuzzy DEMATEL. *International Journal of Applied Decision Sciences*, 7(3), pp. 270–294.

Montazer, G.A., Saremi, H.Q., and Ramezani, M. (2009) Design a new mixed expert decision aiding system using fuzzy ELECTRE III method for vendor selection. *Expert Systems with Applications*, 36(8), pp. 10837–10847. doi: 10.1016/j.eswa.2009.01.019

Mukherjee, S., and Kar, S. (2013) A three phase supplier selection method based on fuzzy preference degree. *Journal of King Saud University - Computer and Information Sciences*, 25(2), pp. 173–185. http://dx.doi.org/10.1016/j.jksuci.2012.11.001

Müller, J.M., Kiel, D., and Voigt, K.-I. (2018) What drives the implementation of Industry 4.0? The role of opportunities and challenges in the context of sustainability. *Sustainability*, 10(1), pp. 659–670. doi: 10.3390/su10010247

Nazari-Shirkouhi, S., Shakouri, H., Javadi, B., and Keramati, A. (2013) Supplier selection and order allocation problem using a two-phase fuzzy multi-objective linear programming. *Applied Mathematical Modelling*, 37(22), pp. 9308–9323. doi: 10.1016/j.apm.2013.04.045

Neumüller, C., Lasch, R., and Kellner, F. (2016) Integrating sustainability into strategic supplier portfolio selection. *Management Decision*, 54(1), pp. 194–221.

Omurca, S.I. (2013) An intelligent supplier evaluation, selection and development system. *Applied Soft Computing Journal*, 13(1), pp. 690–697. doi: 10.1016/j.asoc.2012.08.008

Orji, I.J., and Wei, S. (2015) An innovative integration of fuzzy-logic and systems dynamics in sustainable supplier selection: a case on manufacturing industry. *Computers and Industrial Engineering*, 88, pp. 1–12. doi: 10.1016/j.cie.2015.06.019

Osiro, L., Lima-Junior, F.R., and Carpinetti, L.C.R. (2014) A fuzzy logic approach to supplier evaluation for development. *International Journal of Production Economics*, 153, pp. 95–112. doi: 10.1016/j.ijpe.2014.02.009

Pai, F.-Y., Yeh, T.-M., and Tang, C.-Y. (2018) Classifying restaurant service quality attributes by using Kano model and IPA approach. *Total Quality Management & Business Excellence*, 29(3–4), pp. 301–328. doi: 10.1080/14783363.2016.1184082

Peng, J. (2012) Selection of logistics outsourcing service suppliers based on AHP. *Energy Procedia*, 17, pp. 595–601. doi: 10.1016/j.egypro.2012.02.141

Qian, L. (2014) Market-based supplier selection with price, delivery time, and service level dependent demand. *International Journal of Production Economics*, 147, pp. 697–706. doi: 10.1016/j.ijpe.2013.07.015

Rajesh, G., and Malliga, P. (2013) Supplier selection based on AHP QFD methodology. *Procedia Engineering* 64, pp. 1283–1292. doi: 10.1016/j.proeng.2013.09.209

Rajnai, Z., and Kocsis, I. (2017) Labor market risks of industry 4.0, digitization, robots and AI. In *2017 IEEE 15th International Symposium on Intelligent Systems and Informatics (SISY)*, pp. 000343–000346. doi: 10.1109/SISY.2017.8080580

Rao, C., Goh, M., and Zheng, J. (2017) Decision mechanism for supplier selection under sustainability. *International Journal of Information Technology & Decision Making*, 16(1), pp. 87–115. doi: 10.1142/S0219622016500450

Rouyendegh, B.D., and Saputro, T.E. (2014) Supplier selection using integrated fuzzy TOPSIS and MCGP: a case study. *Procedia-Social and Behavioral Sciences*, 116, pp. 3957–3970.

Ruiz-Torres, A.J., Mahmoodi, F., and Zeng, A.Z. (2013) Supplier selection model with contingency planning for supplier failures. *Computers and Industrial Engineering*, 66(2), pp. 374–382. doi: 10.1016/j.cie.2013.06.021

Sanayei, A., Farid Mousavi, S., Abdi, M.R., and Mohaghar, A. (2008) An integrated group decision-making process for supplier selection and order allocation using multi-attribute utility theory and linear programming. *Journal of the Franklin Institute*, 345(7), pp. 731–747. doi: 10.1016/j.jfranklin.2008.03.005

Sanayei, A., Farid Mousavi, S., and Yazdankhah, A. (2010) Group decision making process for supplier selection with VIKOR under fuzzy environment. *Expert Systems with Applications*, 37(1), pp. 24–30. doi: 10.1016/j.eswa.2009.04.063

Shafia, M.A., and Abdollahzadeh, S. (2014) Integrating fuzzy kano and fuzzy TOPSIS for classification of functional requirements in national standardization system. *Arabian Journal for Science and Engineering*, 39(8), pp. 6555–6565.

Shaik, M., and Abdul-Kader, W. (2011) Green supplier selection generic framework: a multi-attribute utility theory approach. *International Journal of Sustainable Engineering*, 4(1), pp. 37–56. doi: 10.1080/19397038.2010.542836

Shen, C.Y., and Yu, K.T. (2013) Strategic vender selection criteria. *Procedia Computer Science* 17, pp. 350–356. doi: 10.1016/j.procs.2013.05.045

Shen, C., and Yu, K. (2009) Enhancing the efficacy of supplier selection decision-making on the initial stage of new product development: a hybrid fuzzy approach considering the strategic and operational factors simultaneously. *Expert Systems with Applications*, 36(8), pp. 11271–11281. doi: 10.1016/j.eswa.2009.02.083

Shen, L., Olfat, L., Govindan, K., Khodaverdi, R., and Diabat, A. (2013) Resources, conservation and recycling a fuzzy multi criteria approach for evaluating green supplier's performance in green supply chain with linguistic preferences. *Resources, Conservation & Recycling*, 74, pp. 170–179. doi: 10.1016/j.resconrec.2012.09.006

Spangenberg, J.H. (2004) Reconciling sustainability and growth: criteria, indicators, policies. *Sustainable Development*, 12(2), pp. 74–86. doi: 10.1002/sd.229

Stock, T., and Seliger, G. (2016) Opportunities of sustainable manufacturing in Industry 4.0. *Procedia CIRP*, 40, pp. 536–541. doi: 10.1016/j.procir.2016.01.129

Tavana, M., Yazdani, M., and Caprio, D. Di. (2016) An application of an integrated ANP – QFD framework for sustainable supplier selection. *International Journal of Logistics Research and Applications*, 20(3), pp. 254–275. doi: 10.1080/13675567.2016.1219702

Teixeira, A.A., Jabbour, C.J.C., De Sousa Jabbour, A.B.L., Latan, H., and De Oliveira, J.H.C. (2016) Green training and green supply chain management: evidence from Brazilian firms. *Journal of Cleaner Production*, 116, pp. 170–176. doi: 10.1016/j.jclepro.2015.12.061

Thiruchelvam, S., and Tookey, J.E. (2011) Evolving trends of supplier selection criteria and methods. *International Journal of Automotive and Mechanical Engineering*, 4(1), pp. 437–454.

Thoben, K.-D., Wiesner, S., and Wuest, T. (2017) "Industrie 4.0" and smart manufacturing – A review of research issues and application examples. *International Journal of Automation Technology*, 11(1), pp. 4–16. doi: 10.20965/ijat.2017.p0004

Uygun, Ö., and Dede, A. (2016) Performance evaluation of green supply chain management using integrated fuzzy multi-criteria decision making techniques. *Computers & Industrial Engineering*, 102, pp. 502–511. doi: 10.1016/j.cie.2016.02.020

Vinodh, S., Jayakrishna, K., and Girubha, R.J. (2013) Sustainability assessment of an automotive organisation using fuzzy Kano's model. *International Journal of Sustainable Engineering*, 6(1), pp. 1–9. doi: 10.1080/19397038.2011.654362

Waibel, M.W., Steenkamp, L.P., Moloko, N., and Oosthuizen, G.A. (2017) Investigating the effects of smart production systems on sustainability elements, 8, pp. 731–737. *Procedia Manufacturing*. doi: 10.1016/j.promfg.2017.02.094.

Wang, C.-H. (2013) Incorporating customer satisfaction into the decision- making process of product configuration: a fuzzy Kano perspective Incorporating customer satisfaction into the decision-making process of product configuration: a fuzzy Kano perspective. *International Journal of Production Research* 51(22), (August 2014), pp. 6651– 6662. doi: 10.1080/00207543.2013.825742

Wang, C.-H., and Fong, H.-Y. (2016) Integrating fuzzy Kano model with importance-performance analysis to identify the key determinants of customer retention for airline services. *Journal of Industrial and Production Engineering*, 33(7), pp. 450–458. doi: 10.1080/21681015.2016.1155668

Wang, C.-H., and Wang, J. (2014) Combining fuzzy AHP and fuzzy Kano to optimize product varieties for smart cameras: a zero-one integer programming perspective. *Applied Soft Computing Journal*, 22, pp. 410–416. doi: 10.1016/j.asoc.2014.04.013

Wang, W.P. (2010) A fuzzy linguistic computing approach to supplier evaluation. *Applied Mathematical Modelling*, 34(10), pp. 3130–3141. doi: 10.1016/j.apm.2010.02.002

Ware, N.R., Singh, S.P., and Banwet, D.K. (2014) A mixed-integer non-linear program to model dynamic supplier selection problem. *Expert Systems with Applications*, 41(2), pp. 671–678. doi: 10.1016/j.eswa.2013.07.092

Winter, S., and Lasch, R. (2016) Environmental and social criteria in supplier evaluation – lessons from the fashion and apparel industry. *Journal of Cleaner Production*, 139, pp. 175–190. doi: 10.1016/j.jclepro.2016.07.201

Wu, Y., K. Chen, B. Zeng, H. Xu, and Y. Yang (2016) Supplier selection in nuclear power industry with extended VIKOR method under linguistic information. *Applied Soft Computing Journal*, 48, pp. 444–457. doi: 10.1016/j.asoc.2016.07.023

Xiao, Z., W. Chen, and L. Li (2012) An integrated FCM and fuzzy soft set for supplier selection problem based on risk evaluation. *Applied Mathematical Modelling*, 36(4), pp. 1444–1454. doi: 10.1016/j.apm.2011.09.038

Xu, X. (2012) From cloud computing to cloud manufacturing. *Robotics and Computer-Integrated Manufacturing*, 28(1), pp. 75–86. doi: 10.1016/j.rcim.2011.07.002

Yin, S., and Kaynak, O. (2015) Big Data for modern industry: challenges and trends. *Proceedings of the IEEE*, 103(2), pp. 143–146. doi: 10.1109/JPROC.2015.2388958

Yang, S., M.R. Aravind Raghavendra, J. Kaminski, and H. Pepin (2018) Opportunities for Industry 4.0 to support remanufacturing. *Applied Sciences*, 8(7), Art. no. 1177. doi: 10.3390/app8071177.

Zhong, R.Y., Xu, X., Klotz, E., and Newman, S.T. (2017) Intelligent manufacturing in the context of Industry 4.0: a review. *Engineering*, 3(5), pp. 616–630. doi: 10.1016/J.ENG.2017.05.015

Zimmer, K., Fröhling, M., and Schultmann, F. (2016) Sustainable supplier management – a review of models supporting sustainable supplier selection, monitoring and development. *International Journal of Production Research*, 54(5), pp. 1412–1442. doi: 10.1080/00207543.2015.1079340

Zolghadri, M., Eckert, C., Zouggar, S., and Girard, P. (2011) Power-based supplier selection in product development projects. *Computers in Industry*, 62(5), pp. 487–500. doi: 10.1016/j.compind.2010.12.001

Zouggari, A., and Benyoucef, L. (2012) Simulation based fuzzy TOPSIS approach for group multi-criteria supplier selection problem. *Engineering Applications of Artificial Intelligence*, 25(3), pp. 507–519. doi: 10.1016/j.engappai.2011.10.012

4 Supply Chain and the Sustainability Management

Selection of Suppliers for Sustainable Operations in the Manufacturing Industry

Sunil Kumar Jauhar[1], Millie Pant[2], Vimal Kumar[3], Nagendra Kumar Sharma[4], and Pratima Verma[5]

[1]Operations Management and Decision Sciences, Indian Institute of Management Kashipur, India
[2]Applied Science and Engineering, Indian Institute of Technology, Roorkee, 247667, India
[3]Department of Information Management, Chaoyang University of Technology, Taichung-41349, Taiwan
[4]Department of Business Administration, Chaoyang University of Technology, Taichung-41349, Taiwan
[5]Department of Information Management, Chaoyang University of Technology, Taichung-41349, Taiwan

4.1 INTRODUCTION

In organizations, whether private or public, there is a sincere concern about sustainable development and sustainability issues. In every activity, the focus is on developing and maintaining an efficient system so that resources can be used in the present or the future by providing a better safeguard to the environment (Ribeiro and Aibar-Guzman, 2010). Several environmental programs and standards are continuously engaged in environmental protection for upcoming generations (Chavan, 2005). In the context of SCM, the primary focus is on` creating and developing effective supply chain models that are fruitful in terms of profit and have a positive impact on society. Supplier selection, an essential aspect of SCM, is the focus of this chapter. In the past few decades, selecting appropriate suppliers has been a crucial aspect of several functions of supply-chain practices (Carter and Narasimhan, 1996; Spekman et al., 1999).

DOI: 10.1201/9781003102304-4

Besides suppliers' evaluation and development, various roles are highlighted, such as supplier management for quality maintenance, maintaining a relationship for strategic development in the long term via appropriate supplier selection, cutting the supplier base, and other supplier development programs (Modi and Mabert, 2007; Prahinski and Benton, 2004). Since external suppliers supply significant products, it is imperative to focus on the several processes and steps involved in the supply chain by selecting ecological or sustainable suppliers. In today's scenario, decision-makers are supposed to work on conventional parameters such as lead-time controlling, quality controlling, cost control, and higher flexibility by considering environmental challenges. This particular phenomenon in the supply chain area gives insights into novel research tracks, such as SSS. The SSS is not a unique model, but it addresses conventional supplier selection problems by incorporating environmental aspects with profit for suppliers' performance assessment (Genovese et al., 2010). Delivery within time, reliability, price controlling, speed, lead-time management, greenhouse effect, CO_2 emission, product reusability, and quality control, etc., are some of the significant SSS features. The full list of all abbreviations is provided in Appendix A.

In this chapter, an effort is made to develop an efficient and effective SSS system with the VIKOR method's help by incorporating the conventional multi-criteria tool for performance evaluation. The proposed decision-making approach uses the VIKOR method to measure suppliers' efficiency scores in the manufacturing industry. The application of the proposed decision-making approach is illustrated in a hypothetical, but realistic, automobile industry. This chapter is organized into seven sections. The rest of this study is organized as follows. Section 4.2 provides a brief literature review of sustainable supplier selection. Section 4.3 is devoted to the gaps in the literature. Section 4.4 discusses the methodology used, and section 4.5 provides the result and discussion of the current study. Lastly, we provide conclusions in section 4.6.

4.2 SUSTAINABLE SUPPLIER SELECTION (SSS) – BACKGROUND OF THE STUDY AND THEORETICAL FRAMEWORK

There have been several studies on sustainability and selection of suppliers; therefore, various models exist in SSS. In this section, few significant studies are highlighted with new and traditional approaches. In the study by Noci, (1997), the supplier was evaluated based on environmental performance to design a sustainable vendor rating system. The study by Handfield et al. (2002) applied the hierarchical analytical process (AHP) to evaluate and estimate the suppliers' performance in context to environmental scale. On the other hand, Hsu and Hu (2009) worked on the ANP to develop a sustainable supply chain model. The study conducted by Beske et al. (2008) has incorporated social and ecological standards developed by German first-tier Volkswagen AG suppliers. A comparison-based investigation on electronic industries from America, Japan, and Taiwan was conducted for SSS in the study (Chiou et al., 2008). Bai and Sarkis (2010a) applied a rough set theory for SSS in their study. Genovese et al. (2010) conducted a critical review based on SSS. In Grisi et al. (2010) the supplier's

TABLE 4.1
Summary of Individual Approaches

Approaches	Authors
AHP	Noci (1997), Handfield et al. (2002), Chiou et al. (2008), Grisi et al. (2010), Lu et al. (2007), Lee et al. (2009)
ANP	Hsu and Hu (2009), Büyüközkan and Çifçi (2011), Hsu and Hu (2007), Büyüközkan and Çifçi (2011)
Chi-square test	Vachon and Klassen (2006)
Choquet integral	Feyziogelu and Büyüközkan (2010)
Confirmatory factor analysis	Chiou et al. (2011)
Cross case analysis	Bala et al. (2008)
DEA	Kumar and Jain (2010)
DEMATEL	Hsu et al. (2013)
Extensible synthetic model	Yang and Wu (2008)
Factor analysis	Hong-Jun and Bin (2010)
Fuzzy Axiomatic Design	Kannan et al. (2015)
Fuzzy inference method	Humphreys et al. (2006)
Fuzzy multi-agent decision-making model	Zhang et al. (2003)
Fuzzy TOPSIS	Awasthi et al. (2010), Kannan et al. (2014)
Grey entropy synthetic evaluation model	Yang and Wu (2008)
Knowledge-based system	Humphreys et al. (2006)
MOGA	Yeh and Chuang (2011)
Rough set theory	Bai and Sarkis (2010a), Bai and Sarkis (2010b)
Structural equation model	Large and Thomsen (2011)

performance assessment was conducted based on a green supply chain. Feyziogelu and Büyüközkan (2010) discussed sustainable suppliers' assessment based on decision criteria incorporating dependencies. Table 4.1 shows that some of the commonly applied SSS approaches are shown, and Table 4.2 shows that integrated SSS approaches are highlighted.

Apart from the standard performance assessment methods, such as ANP, AHP, and other mathematical measures such as rough set theory, metaheuristics are also applied in the SSS process. Lu et al. (2007) studied a multi-objective modelling approach towards sustainable principles for application to SSS. Lee et al. (2009) showed the SSS decisive model in big industries. They applied fuzzy logic by incorporating AHP. Büyüközkan and Çifçi (2011) have shown a multi-criteria decision framework for SSS based on a fuzzy logic-based approach. In Table 4.3, literature based on multiple criteria modelling towards SSS assessment is summarized.

Yeh and Chuang (2011) studied an exciting Multi-Objective Genetic Algorithm (MOGA) application, which is used to select partners in tackling green supply chain challenges. Kumar et al. (2016) have proposed selecting green suppliers using

TABLE 4.2
Summaries of Integrated Approaches

Approaches	Authors
AHP and ANN	Thongchattu and Siripokapirom (2010)
AHP and GA	Yan (2009)
AHP and Taguchi loss functions	Sivakumar et al. (2015)
ANN-MADA	Kuo et al. (2010)
ANP and Grey relational analysis	Hashemi et al. (2015)
DEA and ANP	Kuo and Lin (2012)
DEA with AHP and ANP	Wen and Chi (2010)
DEA with GA and immune strategy	Kumar et al. (2016)
DEA and genetic programming	Fallahpour et al. (2016)
DEA and goal programming	Shabanpour et al. (2017)
Fuzzy Set Theory and Grey Relational Analysis	Chen et al. (2010)
Grey Correlation Analysis and AHP	Li and Zhao (2009)
Knowledge-based system and Case-based reasoning	Humphreys et al. (2003), Humphreys et al. (2003)

genetic and immune strategies. Fallahpour et al. (2016) have shown the DEA and GP approach towards SSS under the fuzzy approach. Kuo et al. (2010) have worked on two approaches by combining ANN and MADA.

4.3   RESEARCH GAPS

The related literature review shows that increasing attention has been given to sustainable supplier selection problems. However, there is still a need for holistic supplier selection approaches considering the ecological benefits and making important decisions like supplier selection. Further, it was also noted that the performance evaluation of sustainable suppliers, especially in the Indian manufacturing industry, has not been addressed in the literature. The present research is a small contribution towards sustainable supplied selection in the context of the Indian scenario.

The proposed decision-making framework utilizes VIKOR, an MCDM method relatively less explored, for sustainable supplier selection, while considering a crucial environmental factor, i.e., carbon emission.

4.3.1   Contributions to the Study

This study helps in attaining the following objectives:

1. Developing a holistic framework for SSS by considering economic as well as environmental factors.

TABLE 4.3

Summary of Literature on Multiple Criteria Modeling and SSS

Authors	Approach	Application
Noci (1997)	AHP	Automobile Industry
Handfield et al. (2002)	AHP	Automotive, Apparel and Paper Manufacturer
Humphreys et al. (2003)	Knowledge-based system and Case-based reasoning	Telecommunication Company
Zhang et al. (2003)	Fuzzy Multi-Agent Decision Making Model	
Vachon and Klassen (2006)	Chi-Square Test	Package printing Industry
Lu et al. (2007)	FAHP	Electronics Industry
Yang and Wu (2008)	Extensible Synthetic Model	Electrical appliances manufacturing
Chiou et al. (2008)	FAHP	Electronic Industry
Bala et al. (2008)	Cross Case Analysis	University Application
Lee et al. (2009)	FEAHP	Electronics Industry
Hsu and Hu (2009)	ANP	Electronics Industry
Li and Zhao (2009)	Grey Correlational Analysis and AHP	Electronics Company
Yan (2009)	AHP and GA	
Kuo et al. (2010)	ANN-MADA	Digital Camera Manufacturing Company
Chen et al. (2010)	Fuzzy Set Theory and Grey Relational Analysis	Electronics Industry
Hong-Jun and Bin (2010)	Factor Analysis	Manufacturing Industry
Awasthi et al. (2010)	Fuzzy Topics	Logistics
Bai and Sarkis (2010a)	Rough Set Theory	Hypothetical Case
Thongchattu and Siripokapirom (2010)	AHP and ANN	
Wen and Chi (2010)	DEA with AHP and ANP	
Yeh and Chuang (2011)	MOGA	Electronics Industry
Kuo et al. (2010)	DEA and ANP	High-tech Industry
Büyüközkan and Çifçi (2011)	FANP	Manufacturing Industry
Hsu et al. (2013)	DEMATEL	Electronics Industry
Kannan et al. (2014)	Fuzzy TOPSIS	electronics company
Kumar et al. (2014)	DEA	Automaker industry
Kannan et al. (2015)	Fuzzy Axiomatic Design	electronics company
Hashemi et al. (2015)	ANP and Grey relational analysis	Automotive industry
Sivakumar et al. (2015)	AHP and Taguchi loss functions	Mining industry
Kumar et al. (2016)	DEA and genetic/immune strategy	Automaker industry
Fallahpour et al. (2016)	DEA and genetic programming	Garment manufacturer
Shabanpour et al. (2017)	DEA and goal programming	Engineering, procurement, and construction

2. Performing performance measurement for a sustainable supplier through VIKOR, and providing recommendations for each supplier/partner to improve its performance, ultimately leading to the improvement of the total sustainable supply chain performance;

3. Using India-based manufacturing industry for application;

4. Analyzing and conferring managerial insights.

4.4 METHODOLOGY

As already mentioned, this study aims to evaluate the performance management and optimization of the supply chain by managing its sustainable supplier performance in the manufacturing industry. For this purpose, the VIKOR ("*VlseKriterijumska Optimizacija I Kompromisno Resenje*" translated as "*multi-criteria optimization and compromise solution*") method is applied as a multi-criteria decision-making method (MCDM). This method is beneficial when experts and decision-makers cannot express their preferences at the beginning of the decision-making process.

VIKOR is a simple method, easy to understand, and convenient to apply. It provides a standardized and convincing approach to solve the problems and produce the best decision. Compared to other MCDM methods, VIKOR provides an overall comparison score that aggregate all criteria and their measurement together. Moreover, this method's compromise solution can be the foundation for negotiations, including decision-makers' preference of criteria weights.

VIKOR serves as an appropriate tool for measuring supply chain operations performance for determining efficient and green suppliers. For more real-world VIKOR method applications, the interested reader may refer to (Opricovic and Tzeng, 2004; 2007). VIKOR is regarded as one of the most successful efficiency assessment techniques internationally by researchers in management science and operations research.

This study presents the research objective to evaluate the performance management and optimization of the supply chain by managing its sustainable supplier performance in the manufacturing industry. The input/output criteria utilized for showing the applicability of VIKOR for SSS are taken from existing literature. Jauhar and Pant (2017) and Shirouyehzad et al. (2009) have identified Lead Time (L), Qualities (Q), and Price (P) as input criteria and Service quality (S.Q.), and CO_2 emission (C.E.) as output criteria. After a brainstorming session with experts from industry and academia, eighteen suppliers and five input/output criteria were identified for this study, followed by prioritizing the appropriate suppliers and their performance to the manufacturing industry using VIKOR. The flowchart of the study has been shown in Figure 4.1.

VIKOR was introduced as a multi-criteria decision-making tool (Opricovic and Tzeng, 2004; 2007). The fundamental concept of VIKOR lies in exploring the "*negative and positive the ideal solutions*", identifying the alternatives with the lowest (negative) and the highest (positive) values. VIKOR is a multi-criteria ranking tool for measuring the ideal solution's proximity through a linear standardization procedure. After that, a compromise solution is identified as the best

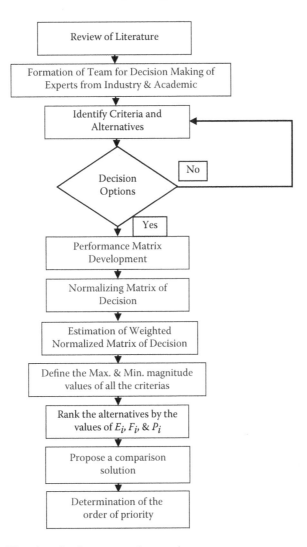

FIGURE 4.1   Flowchart for the suggested research concept.

alternative from a set of feasible alternatives with conflicting criteria. It has given the maximum cluster value for the "majority" and the minimum distinct guilt for the "rival".

The analysis-based solution denotes a viable solution, nearby to the optimistic perfect solution and utmost from the adverse idyllic solution. $L_{1, i}$ and $L_{\infty,i}$ are used to formulate the ranking measure. The necessary steps involved in applying VIKOR can be seen (Opricovic and Tzeng, 2004; 2007). In the beginning, a decision matrix is generated, and the best, $(x_{ij})_{max}$ and the worst, $(x_{ij})_{min}$ values for all the criteria are determined as E_i, F_i, and P_i.

$E_{i\,max}$, $E_{i\,min}$, $F_{i\,max}$, $F_{i\,min}$ represent the maximum and minimum values. Experts and decision-makers locate the measurement value of v in the range 0 and 1 as *"the majority of attributes"*. Three scenarios with $v > 0.5$ (first), $v < 0.5$ (second), and $v = 0.5$ (third) are considered. The first case represents the primary word in the given Equation (4.5). Therefore, to the substitutes worldwide presentation in respect to the whole of the criteria, the second case is vice versa of first while the third case would be considered once both of these perspectives are taken into consideration in a balanced way. In the next step, the finest substitute is measured as having the lowest Pi of value arranged in ascending order. The necessary parameters are calculated through the Equations (4.1–4.5):

$$L_{p,i} = \left\{ \sum_{j=1}^{n} \left(w_j \frac{[(x_{ij})_{max} - x_{ij}]}{[(x_{ij})_{max} - (x_{ij})_{min}]} \right)^p \right\}^{1/p} \quad 1 \leq p \leq \infty; \ i = 1, 2, ..., m \quad (4.1)$$

$$E_i = L_{p,i} = \sum_{j=1}^{n} w_j \frac{[(x_{ij})_{max} - x_{ij}]}{[(x_{ij})_{max} - (x_{ij})_{min}]} \quad (4.2)$$

$$F_i = L_{\infty,i} = Max^m of \left\{ w_j \frac{[(x_{ij})_{max} - x_{ij}]}{[(x_{ij})_{max} - (x_{ij})_{min}]} \right\} j = 1, 2, ,..., n \quad (4.3)$$

$$E_i = L_{1,i} = \sum_{j=1}^{n} w_j \frac{[x_{ij} - (x_{ij})_{min}]}{[(x_{ij})_{max} - (x_{ij})_{min}]} \quad (4.4)$$

$$P_i = v \left(\frac{[E_i - (E_i)_{min}]}{[(E_i)_{max} - (E_i)_{min}]} \right) + (1 - v) \left(\frac{[F_i - (F_i)_{min}]}{[(F_i)_{max} - (F_i)_{min}]} \right) \quad (4.5)$$

4.5 RESULT AND DISCUSSION

The study has highlighted VIKOR steps based on the proposed flowchart so that the rankings can be obtained for all 18 performances of the suppliers chosen for effective decision-making. The authors opted for these input and output criteria from Opricovic and Tzeng (2004). The study also endeavours to develop a model that can be better used for prioritizing the selection of sustainable suppliers through VIKOR. The results based on the analysis and prioritizing the given variables help make decisions, both strategic and tactical. These decisions are further significant for Indian manufacturing industries in sustainable operational practices.

TABLE 4.4

The Marks or the Scores Based on the Matrix of Decision and Weights Received by Various Experts and Specialist's Recommendation Between Best and Worst Values for Each Criterion (for Input Criteria), Note: NB: Non-Beneficial Criteria and B: Beneficial Criteria

SN	Weight →	0.3	0.45	0.25
	Attribute/Criteria	L	Q	P
	Supplier	NB	B	NB
1	SUP_1	3	75	187
2	SUP_2	2	77	195
3	SUP_3	4	85	272
4	SUP_4	5	86	236
5	SUP_5	3	74	287
6	SUP_6	6	62	242
7	SUP_7	3	73	168
8	SUP_8	4	92	396
9	SUP_9	2	77	144
10	SUP_10	3	69	137
11	SUP_11	6	54	142
12	SUP_12	5	57	196
13	SUP_13	1	77	247
14	SUP_14	3	61	148
15	SUP_15	4	69	294
16	SUP_16	6	94	249
17	SUP_17	2	88	121
18	SUP_18	1	78	269
	Best $[(x_{ij})_{max}]$	1	94	121
	Worst $[(x_{ij})_{min}]$	6	54	396

This model's execution, including top to lowermost ranking and suppliers, provides greater focus, as showcased in the present part. So, on the basis of tactical based alternatives of five makers of decision, the judgment is significant since they have provided the weight of "0.3" "0.45", and "0.25" for input criteria such as Lead Time (L), Qualities (Q), and Price (P), respectively. The criteria weight for Service quality (S.Q.) and CO2 emission (C.E.) is given as "0.4" and "0.6", respectively. These eighteen suppliers are given a score to input and output criteria (Opricovic and Tzeng, 2004). Further, these criteria are categorized into non-beneficial and beneficial, with their scores and relative weights given in Table 4.4 (for input data) and Table 4.7 (for input data) considering methods step 1 and 2.

TABLE 4.5

The Normalized Decision Matrix With Minimum and Maximum Values (for Input Criteria). *Note: $E_{i(min)}$=Min of E_i, $F_{i(min)}$= Min of F_i, $E_{i(max)}$= Max of E_i, F_i $_{(max)}$= Max of F_i*

SN	Attribute/Criteria	L	Q	P	E_i	F_i
1	SUP_1	0.12	0.214	0.06	0.394	0.214
2	SUP_2	0.06	0.191	0.067	0.319	0.191
3	SUP_3	0.18	0.101	0.137	0.419	0.18
4	SUP_4	0.24	0.09	0.105	0.435	0.24
5	SUP_5	0.12	0.225	0.151	0.496	0.225
6	SUP_6	0.3	0.360	0.110	0.770	0.360
7	SUP_7	0.12	0.236	0.043	0.399	0.236
8	SUP_8	0.18	0.023	0.250	0.453	0.250
9	SUP_9	0.06	0.191	0.021	0.272	0.191
10	SUP_10	0.12	0.281	0.015	0.416	0.281
11	SUP_11	0.3	0.450	0.019	0.769	0.450
12	SUP_12	0.24	0.416	0.068	0.724	0.416
13	SUP_13	0	0.191	0.115	0.306	0.191
14	SUP_14	0.12	0.371	0.025	0.516	0.371
15	SUP_15	0.18	0.281	0.157	0.619	0.281
16	SUP_16	0.3	0	0.116	0.416	0.300
17	SUP_17	0.06	0.068	0.00	0.128	0.068
18	SUP_18	0	0.180	0.135	0.315	0.18
	$E_{i(min)}$, $F_{i(min)}$				0.128	0.068
	$E_{i(max)}$, $F_{i(max)}$				0.770	0.450

The VIKOR technique and its steps are adopted in the study and shown in Tables 4.4 to 4.9. The normalized decision matrix with minimum and maximum values, given in Table 4.5 (for input data) and Table 4.8 (for input data) considering method step 3. Based on these weighted normalized decision matrices, we evaluate all criteria' maximum and minimum magnitude values with the value of the weight of the strategy v, the maximum group utility here. Using step 4, we find the rank of alternatives by Ei, Fi, and Pi values. Low values of Pi represent the relatively higher rank. Table 4.6 (for input data) and Table 4.9 (for input data) represent the summary of the final selection and ranking scores considering step 4.

The final rank of the selection of suppliers for input criteria is given as S17>S9>S18>S13>S2>S3>S1>S7>S4>S8>S5>S10>S16>S15>S14>S6>S12>-S11. Table 4.6 shows that S17 and S11 have the lowest and highest Pi scores, respectively, for input criteria. The final rank of the selection of suppliers for output criteria is given as S15>S6>S16>S11>S17>S7>S10>S2>S12>S9>S14>

TABLE 4.6
Summary of the Scores for Final Selection and Ranking (for Input Criteria)

SN	Attribute/Criteria	E_i	F_i	P_i	Rank based on P_i
1	SUP_1	0.394	0.214	0.399	7
2	SUP_2	0.319	0.191	0.310	5
3	SUP_3	0.419	0.18	0.374	6
4	SUP_4	0.435	0.24	0.464	9
5	SUP_5	0.496	0.225	0.493	11
6	SUP_6	0.770	0.360	0.882	16
7	SUP_7	0.399	0.236	0.439	8
8	SUP_8	0.453	0.250	0.491	10
9	SUP_9	0.272	0.191	0.274	2
10	SUP_10	0.416	0.281	0.504	12
11	SUP_11	0.769	0.450	0.999	18
12	SUP_12	0.724	0.416	0.920	17
13	SUP_13	0.306	0.191	0.301	4
14	SUP_14	0.516	0.371	0.699	15
15	SUP_15	0.619	0.281	0.662	14
16	SUP_16	0.416	0.300	0.529	13
17	SUP_17	0.128	0.068	0.000	1
18	SUP_18	0.315	0.18	0.293	3

S13>S5>S1>S8>S3>S4>S18. Table 4.9 shows that S15 and S18 have the lowest and highest Pi scores, respectively, for output criteria. The graphs have been plotted for the priority of input and output criteria in Figure 4.2 and Figure 4.3, respectively. These graphs show the final rank of input and output suppliers selection for the manufacturing units. These suppliers are identified as the key suppliers of their performance for sustainable operations in the manufacturing industry. Their prioritization of this supplier selection recommends the current manufacturing units so that their sequence can be chosen properly, and additionally, their competitive performance can be increased significantly for survival in the current competitive business environment.

4.6 CONCLUSION

The present study contributes to the development and encouragement of sustainable supplier selection (SSS), a pertinent issue in today's business environment. For any industry, many suppliers are available in the market, and it becomes a tough managerial decision to identify the efficient suppliers from an economic point of view and are efficient from the viewpoint of the environment. With the advent of efficient decision-making tools that can deal with multiple criteria simultaneously, the extra burden of

TABLE 4.7

The Scores or Marks Matrix of Decision and Weights by Specialists and Various Experts' Recommendation With Best and Worst Values for Each Criterion (for Output Criteria), Note: NB: Non-Beneficial Criteria and B: Beneficial Criteria

SN	Weight →	0.4	0.6
	Attribute/Criteria	S.Q.	C.E.
	Supplier	B	NB
1	SUP_1	84	40
2	SUP_2	76	30
3	SUP_3	27	25
4	SUP_4	110	22
5	SUP_5	94	38
6	SUP_6	102	10
7	SUP_7	82	24
8	SUP_8	63	38
9	SUP_9	55	26
10	SUP_10	61	18
11	SUP_11	122	24
12	SUP_12	75	30
13	SUP_13	80	55
14	SUP_14	121	39
15	SUP_15	125	8
16	SUP_16	76	6
17	SUP_17	114	55
18	SUP_18	65	48
	Best $[(x_{ij})_{max}]$	125	6
	Worst $[(x_{ij})_{min}]$	27	55

selecting SSS has reduced to some extent for the managers or top decision-making bodies of an industry. Some concluding remarks and future directions that can be drawn from the present study are:

1. This study is quite appropriate from the Indian context. It presents the essential input/output criteria relevant to any manufacturing industry. However, these criteria may be customized based on the type of item being manufactured in an industry.

2. The present study shows the application of VIKOR (a conventional decision-making tool) for SSS; however, as per the current research trend, many new concepts are being integrated into conventional approaches, like VIKOR. Some of these are integrating the concepts of soft computing like

TABLE 4.8

The Normalized Decision Matrix With Minimum and Maximum Values (for Output Criteria), *Note:* $E_{i(min)}$=*Min of* E_i, $F_{i(min)}$= *Min of* F_i, $E_{i(max)}$= *Max of* E_i, $F_{i(max)}$= *Max of* F_i

SN	Attribute/Criteria	S.Q.	SE	E_i	F_i
1	SUP_1	0.167	0.416	0.584	0.416
2	SUP_2	0.200	0.294	0.494	0.294
3	SUP_3	0.400	0.233	0.633	0.400
4	SUP_4	0.510	0.196	0.706	0.510
5	SUP_5	0.127	0.392	0.518	0.392
6	SUP_6	0.094	0.049	0.143	0.094
7	SUP_7	0.176	0.220	0.396	0.220
8	SUP_8	0.253	0.392	0.645	0.392
9	SUP_9	0.286	0.245	0.531	0.286
10	SUP_10	0.261	0.147	0.408	0.261
11	SUP_11	0.012	0.220	0.233	0.220
12	SUP_12	0.204	0.294	0.498	0.294
13	SUP_13	0.184	0.600	0.784	0.184
14	SUP_14	0.016	0.404	0.420	0.404
15	SUP_15	0.000	0.025	0.024	0.024
16	SUP_16	0.200	0.000	0.200	0.200
17	SUP_17	0.045	0.600	0.645	0.045
18	SUP_18	0.245	0.514	0.759	0.514
	$E_{i(min)}$, $F_{i(min)}$			0.024	0.024
	$E_{i(max)}$, $F_{i(max)}$			0.784	0.514

Neural Networks and Fuzzy Logic. Though these two are not exactly new, their potential of being integrated with MCDM methods is relatively less explored.

3. Another interesting direction of research can be integrating the concepts of Machine Learning or Artificial Intelligence into conventional decision-making tools to provide more realistic and applicable solutions

TABLE 4.9

Summary of the Scores for Final Selection and Ranking (for Output Criteria)

SN	Attribute/Criteria	E_i	F_i	P_i	Rank based on P_i
1	SUP_1	0.368	0.400	0.768	14
2	SUP_2	0.309	0.275	0.584	8
3	SUP_3	0.401	0.383	0.784	16
4	SUP_4	0.449	0.496	0.945	17
5	SUP_5	0.325	0.375	0.700	13
6	SUP_6	0.078	0.071	0.149	2
7	SUP_7	0.245	0.200	0.445	6
8	SUP_8	0.409	0.375	0.784	15
9	SUP_9	0.333	0.267	0.600	10
10	SUP_10	0.253	0.242	0.494	7
11	SUP_11	0.137	0.200	0.337	4
12	SUP_12	0.312	0.275	0.587	9
13	SUP_13	0.500	0.163	0.663	12
14	SUP_14	0.261	0.388	0.648	11
15	SUP_15	0.000	0.000	0.000	1
16	SUP_16	0.116	0.179	0.295	3
17	SUP_17	0.409	0.021	0.429	5
18	SUP_18	0.484	0.500	0.984	18

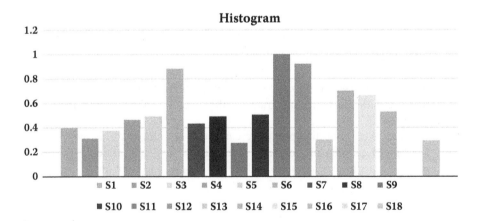

FIGURE 4.2   Rank of the input supplier selection.

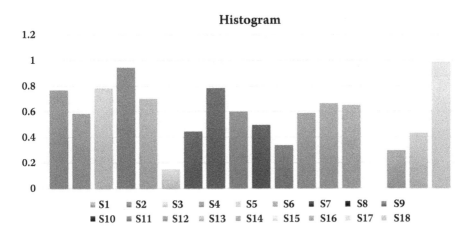

FIGURE 4.3   Rank of the output supplier selection.

REFERENCES

Awasthi, A., Chauhan, S.S., and Goyal, S.K. (2010) A fuzzy multi-criteria approach for evaluating environmental performance of suppliers. *International Journal of Production Economics*, 126, pp. 370–378.

Bai, C., and Sarkis, J. (2010a) Green supplier development: analytical evaluation using rough set theory. *Journal of Cleaner Production*, 18(12), pp. 1200–1210.

Bai, C., and Sarksi, J. (2010b) Integrating sustainability into supplier selection with grey system and rough set methodologies. *International Journal of Production Economics*, 124(1), pp. 252–264.

Bala, A., Paco Muñoz, P., Rieradevall, J., and Ysern, P. (2008) Experiences with greening suppliers. The Universitat Autònoma de Barcelona. *Journal of Cleaner Produc-tion*, 16(15), pp. 1610–1619.

Beske, P., Koplin, J., and Seuring, S. (2008) The use of environmental and social standards by German first-tier suppliers of the Volkswagen AG. *Corporate Social Responsibility and Environmental Management*, 15(2), pp. 63–75.

Büyüközkan, G., and Çifçi, G. (2010) Evaluation of the green supply chain management practices: a fuzzy ANP approach. *Production Planning & Control, iFirst*, pp. 1–14.

Büyüközkan, G., and Çifçi, G. (2011) A novel fuzzy multi-criteria decision framework for sustainable supplier selection with incomplete information. *Computers in Industry*, 62, pp. 164–174.

Carter, J.R., and Narasimhan, R. (1996) Is purchasing really strategic?. *International Journal of Purchasing and Materials Management*, 32(4), pp. 20–28.

Chavan, M. (2005) An appraisal of environment management systems: a competitive ad-vantage for small businesses. *Management of Environmental Quality: An International Journal*, 16(5), pp. 444–463.

Chen, C.C., Tseng, M.L., Lin, Y.H., and Lin, Z.S. (Dec. 2010) Implementation of green supply chain management in uncertainty. In: 2010 IEEE International Conference on Industrial Engineering and Engineering Management (IEEM), pp. 260–264.

Chiou, C.Y., Hsu, C.W., and Hwang, W.Y. (Dec. 2008) Comparative investigation on green supplier Selection of the American, Japanese and Taiwanese electronics industry in China. In: International Conference on IE&EM, IEEE 8e11, pp. 1909–1914.

Chiou, T.Y., Chan, H.K., Lettice, F., and Chung, S.H., (2011) The influence of greening the suppliers and green innovation on environmental performance and competitive advantage in Taiwan. *Transportation Research, Part E*, 47, pp. 822–836.

Fallahpour, A., Olugu, E.U., Musa, S.N., Khezrimotlagh, D., and Wong, K.Y. (2016) An integrated model for green supplier selection under fuzzy environment: application of data envelopment analysis and genetic programming approach. *Neural Computing and Applications*, 27(3), pp. 707–725.

Feyziogelu, O., and Büyüközkan, G. (2010) Evaluation of green suppliers considering decision criteria dependencies. In: *Multiple criteria decision making for sustainable energy and transportation systems, Part 2 (Lecture notes in economics and mathematical systems 1*, vol. 634. pp. 145–154.

Genovese, A., Koh, S.L., Bruno, G., and Bruno, P. (Oct. 2010) Green supplier selection: a literature review and a critical perspective. In: 2010 8th International Conference on Supply Chain Management and Information Systems (SCMIS), pp. 1–6.

Grisi, R.M., Guerra, L., and Naviglio, G. (2010) Supplier performance evaluation for green supply chain management. *Business Performance Measurement and Management*, Part 4, pp. 149–163.

Handfield, R., Walton, S.V., Sroufe, R., and Melnyk, S.A. (2002) Applying environmental criteria to supplier assessment: a study in the application of the analytical hierarchy process. *European Journal of Operational Research*, 141, pp. 70–87.

Hashemi, S.H., Karimi, A., and Tavana, M. (2015) An integrated green supplier selection approach with analytic network process and improved Grey relational analysis. *International Journal of Production Economics*, 159, pp. 178–191.

Hong-Jun, L., and Bin, L., (Mar. 2010) A research on supplier assessment indices system of green purchasing. In: International Conference on ICEE, IEEE 13e14, pp. 314–317.

Hsu, C.W., Kuo, T.C., Chen, S.H., and Hu, A.H. (2013) Using DEMATEL to develop a carbon management model of supplier selection in green supply chain management. *Journal of cleaner production*, 56, pp. 164–172.

Hsu, C.W., and Hu, A.H. (Dec. 2007) Application of analytic network process on supplier selection to hazardous substance management in green supply chain management. In: International Conference on IEEM, IEEE 2e4, pp. 1362–1368.

Hsu, C.W., and Hu, A.H., (2009) Applying hazardous substance management to supplier selection using analytic network process. *Journal of Cleaner Production*, 17, pp. 255–264

Humphreys, P., McCloskey, A., McIvor, R., Maguire, L., and Glackin, C. (2006) Employing dynamic fuzzy membership functions to assess environmental performance in the supplier selection process. *International Journal of Production Research*, 44 (12), pp. 2379–2419.

Humphreys, P., McIvor, R., and Chan, F. (2003) Using case-based reasoning to evaluate supplier environmental management performance. *Expert Systems with Applications*, 25(2), pp. 141–153.

Humphreys, P.K., Wong, Y.K., and Chan, F.T.S. (2003) Integrating environmental criteria into the supplier selection process. *Journal of materials processing technology*, 138(1), pp. 349–356.

Jauhar, S.K., and Pant, M. (2017) Integrating DEA with D.E. and MODE for sustainable supplier selection. *Journal of Computational Science*, 21, pp. 299–306.

Kannan, D., de Sousa Jabbour, A.B.L., and Jabbour, C.J.C. (2014) Selecting green suppliers based on GSCM practices: using fuzzy TOPSIS applied to a Brazilian electronics company. *European Journal of Operational Research*, 233(2), pp. 432–447.

Kannan, D., Govindan, K., and Rajendran, S. (2015) Fuzzy axiomatic design approach based green supplier selection: a case study from Singapore. *Journal of Cleaner Production*, 96, pp. 194–208.

Kumar, A., and Jain, V. (Oct. 2010) Supplier selection: a green approach with carbon footprint monitoring. In: International Conference on SCMIS, IEEE 6e9, pp. 1–9.

Kumar, A., Jain, V., and Kumar, S. (2014) A comprehensive environment friendly approach for supplier selection. *Omega*, 42(1), pp. 109–123.

Kumar, A., Jain, V., Kumar, S., and Chandra, C. (2016) Green supplier selection: a new genetic/immune strategy with industrial application. *Enterprise Information Systems*, 10(8), pp. 911–943.

Kuo, R.J., and Lin, Y.J. (2012) Supplier selection using analytic network process and data envelopment analysis. *International Journal of Production Research*, 50(11), pp. 2852–2863.

Kuo, R.J., Wang, Y.C., and Tien, F.C. (2010) Integration of artificial neural network and MADA methods for green supplier selection. *Journal of cleaner production*, 18(12), pp. 1161–1170.

Large, R.O., and Thomsen, C.G. (2011) Drivers of green supply management performance: evidence from Germany. *Journal of Purchasing and Supply Management*, 17(3), pp. 176–184.

Lee, A.H.I., Kang, H.Y., Hsu, C.F., and Hung, H.C. (2009) A green supplier selection model for high-tech industry. *Expert Systems with Applications*, 36, pp. 7917–7927.

Li, X., and Zhao, C. (Oct. 2009) Selection of suppliers of vehicle components based on green supply chain. In 16th International Conference on Industrial Engineering and Engineering Management, IE&EM'09, pp. 1588–1591.

Lima Ribeiro, V.P., and Aibar-Guzman, C. (2010) Determinants of environmental accounting practices in local entities: evidence from Portugal. *Social Responsibility Journal*, 6(3), pp. 404–419.

Lu, L.Y.Y., Wu, C.H., and Kuo, T.C. (2007) Environmental principles applicable to green supplier evaluation by using multi-objective decision analysis. *International Journal of Production Research*, 45(18–19), pp. 4317–4331.

Modi, S.B., and Mabert, V.A. (2007) Supplier development: improving supplier performance through knowledge transfer. *Journal of Operations Management*, 25(1), pp. 42–64.

Noci, G. (1997) Designing ''green'' vendor rating systems for the assessment of a suppliers environmental performance. *European Journal of Purchasing and Supply Management*, 3(2), pp. 103–114.

Opricovic, S., and Tzeng, G.H. (2004) Compromise solution by MCDM methods: A comparative analysis of VIKOR and TOPSIS. *European Journal of Operational Research*, 156(2), pp. 445–455.

Opricovic, S., and Tzeng, G.H. (2007) Extended VIKOR method in comparison with outranking methods. *European Journal of Operational Research*, 178(2), pp. 514–529.

Prahinski, C., and Benton, W.C. (2004) Supplier evaluations: communication strategies to improve supplier performance. *Journal of Operations Management*, 22(1), pp. 39–62.

Shabanpour, H., Yousefi, S., and Saen, R.F. (2017) Future planning for benchmarking and ranking sustainable suppliers using goal programming and robust double frontiers DEA. *Transportation Research Part D: Transport and Environment*, 50, pp. 129–143.

Shirouyehzad, H., Lotfi, F.H., and Dabestani, R. (Jul. 2009) A data envelopment analysis approach based on the service quality concept for vendor selection. In 2009 International Conference on Computers & Industrial Engineering, pp. 426–430.

Sivakumar, R., Kannan, D., and Murugesan, P. (2015) Green vendor evaluation and selection using AHP and Taguchi loss functions in production outsourcing in mining industry. *Resources Policy*, 46, pp. 64–75.

Spekman, R.E., Kamauff, J., and Spear, J. (1999) Towards more effective sourcing and supplier management. *European Journal of Purchasing & Supply Management*, 5(2), pp. 103–116.

Thongchattu, C., and Siripokapirom, S. (Aug. 2010) Notice of Retraction Green supplier selection consensus by neural network. In 2010 2nd International Conference on Mechanical and Electronics Engineering (ICMEE), pp. V2-313.

Vachon, S., and Klassen, R.D., (2006) Green project partnership in the supply chain: the case of the package printing industry. *Journal of Cleaner Production*, 14, pp. 661–671.

Wen, U.P., and Chi, J.M. (Oct. 2010) Developing green supplier selection procedure: a DEA approach. In 2010 IEEE 17th International Conference on Industrial Engineering and Engineering Management (IE&EM), pp. 70–74.

Yan, G.E. (May 2009) Research on green suppliers' evaluation based on AHP & genetic algorithm. In 2009 International Conference on Signal Processing Systems, pp. 615–619.

Yang, Y., and Wu, L. (Oct. 2008) Extension method for green supplier selection. In: International Conference on WiCom, IEEE 12e14, pp. 1–4.

Yeh, W.C., and Chuang, M.C., (2011) Using multi-objective genetic algorithm for partner selection in green supply chain problems. *Expert Systems with Applications*, 38, pp. 4244–4253

Zhang, H.C., Li, J., and Merchant, M.E. (2003) Using fuzzy multi-agent decision-making in environmentally conscious supplier management. *CIRP Annals - Manufacturing Technology*, 52(1), pp. 385–388.

APPENDIX A

TABLE A
Complete Explanations of the Abbreviations Used in Article

AHP:	*"Analytical Hierarchy Process"*
ANN:	*"Artificial Neural Networks"*
ANP:	*"Analytical Network Process"*
BSC:	*"Balanced scorecard"*
CE:	*"CO_2emission"*
DEA:	*"Data Envelop Analysis"*
DEAP:	*"Data Envelopment Analysis Program"*
DEMATEL:	*"Decision-making Trial and Evaluation Laboratory"*
EMS:	*"Environmental Management Systems"*
FAHP:	*"Fuzzy Analytical Hierarchy Process"*
FANP:	*"Analytical Network Process"*
FEAHP:	*"Fuzzy Extended Analytical Hierarchy Process"*
GAs:	*"Genetic Algorithms"*
GP:	*"Genetic Programming"*
L:	*"Lead Time"*
MADA:	*"Multi Attribute Decision Analysis"*
MCDA:	*"Multi Criteria Decision Analysis"*
MDS:	*"Multidimensional Scaling"*
MLP:	*"Multi Linear Programming"*
MOGA:	*"Multi Objective Genetic Algorithms"*
P:	*"Price"*

TABLE A (Continued)
Complete Explanations of the Abbreviations Used in Article

Q:	*"Qualities"*
SCM:	*"Supply Chain Management"*
SCs:	*"Supply chains"*
SQ:	*"Service Quality"*
SSCM:	*"Sustainable Supply Chain Management"*
SSS:	*"Sustainable Supplier selection"*
SUP:	*"Supplier"*

5 WASPAS Multi-Criteria Decision-Making Approach for Selecting Oxygen Delignification Additives in the Pulp and Paper Industry

Kumar Anupam[1,2], Pankaj Kumar Goley[3], and Anil Yadav[1]

[1]Department of Chemical Engineering, Deenbandhu Chhotu Ram University of Science and Technology, Murthal, Sonipat – 131039, Haryana, India

[2]Chemical Recovery and Biorefinery Division, Central Pulp and Paper Research Institute, Himmat Nagar, Saharanpur – 247001, Uttar Pradesh, India

[3]Engineering and Maintenance Division, Central Pulp and Paper Research Institute, Himmat Nagar, Saharanpur – 247001, Uttar Pradesh, India

5.1 INTRODUCTION

The Industry 4.0 concept pertains to the fourth industrial revolution. It aims to bridge the gap between the real and virtual industrial environments through automation and digitalization by implementing technologies such as IoT (internet of things), CPS (cyber-physical systems), cloud computing, machine learning, AI (artificial intelligence), robotics, additive manufacturing, and advanced data analytics to bring operational transformation and innovation in products, services, and business models for establishing smart industries. The Industry 4.0 concept has also been well received in the pulp and paper industry (Frias et al., 2019). Industry 4.0 relies heavily on the analysis of real-time data to make efficient decisions using simulation, optimization, control, and decision support strategies for processes, products, and business. MCDM (multi-criteria decision-making) methods are employed to solve a diverse range of decision-making problems encountered in various

DOI: 10.1201/9781003102304-5

industrial sectors (Stojčić et al., 2019). There are many critical situations in the pulp and paper sector where MCDM approaches have been applied for making effective decisions.

MCDM methods – viz. Delphi – AHP (analytic hierarchy process), interval BWM (best-worst method), ISWM (interval sum weighted method), interval TOPSIS, fuzzy AHP, TOPSIS – AHP, entropy integrated TOPSIS, entropy integrated SAW, etc. – have also been used to solve several decision-making problems of the pulp and paper industry, such as biorefinery implementation, boxboard production, appropriate pulp-making technology selection, supplier-selection model, raw-material selection for pulp and paper production, coating colour formulation for coated paper production, etc. (Anupam et al., 2014; 2015; Kaushik et al., 2015; Akhundzadeh & Shirazi, 2017; Brunnhofer et al., 2020; Man et al., 2020; Navarro et al., 2020). However, to the best of our knowledge, there is no study available in the literature that reports the application of the WASPAS (weighted aggregated sum product assessment)-based MCDM method for solving any pulp and paper-based selection problems. This chapter focuses on the implementation of the WASPAS method for selecting the most suitable additive to be used for O_2 delignification in the pulp and paper industry.

The WASPAS method is one of the latest MCDM techniques. It was first reported by Zavadskas et al. (2012). This method has emerged as an efficacious decision-making technique in recent years and presents a peculiar consolidation of WSM (weighted sum model) and WPM (weighted product model). The WASPAS method searches for a combined criterion of optimality based on the dual-criteria of optimality wherein the first is derived from WSM while the second is derived from WPM (Chakraborty et al., 2015). WSM and WPM are well-known and well-practised MCDM methods. The accuracy of WASPAS is 1.6 and 1.3 times greater than WSM and WPM, respectively (Kumar et al., 2020). Like other MCDM techniques, the performance of WASPAS is greatly influenced by normalization techniques and criteria-weighting methods. Generally, linear normalization techniques yield better results in the WASPAS method. Different criteria weighting methods such as entropy, SWARA (step-wise weight assessment ratio analysis), AHP, CRITIC (criteria importance through inter criteria correlation), equal weights method, standard deviation method, etc., have been tested for WASPAS. With the progress of research, several versions of WASPAS, such as fuzzy WASPAS, extended WASPAS, rough WASPAS, neutrosophic WASPAS, etc. have evolved. WASPAS finds application in various manufacturing environments (Chakraborty & Zavadskas, 2014).

O_2 delignification is an intermediate process that is executed after pulping but before bleaching stages in the pulp and paper mills. The primary objective of this environmentally benign process is to get rid of the residuary lignin present in the pulp after pulping with the use of O_2 under alkaline conditions to mitigate colour, AOX (adsorbable organic halides), BOD (biological oxygen demand), and COD (chemical oxygen demand) of the discharge effluent (Anupam et al., 2018). However, the severe processing conditions of O_2 delignification (alkali dose, temperature, and oxygen pressure) in contrast cause the degradation of the cellulose leading to poor quality. Therefore, researchers advocate the use of varieties of additives during O_2 delignification of pulp to protect this carbohydrate degradation and increase the O_2 selectivity. Some of them are starch, carboxymethyl cellulose, galactomannan, xylan, glucomannan, ethylene glycol, hydroquinone compounds, sugar-based polymeric compounds, surfactants, chelating agents, tetraacetylethylenediamine, urea, magnesium hydroxide, poly-pyridine, poly-oxometallates, anthraquinone, magnesium sulfate, hydrogen peroxide, transition metal complexes, sodium borohydride, organophosphate, salts, etc. (Bhardwaj et al., 2017). These additives impart different physicochemical properties to pulp and paper, even of the same origin. Often, the properties of O_2 delignified pulp and paper are conflicting in nature. Though the performance of these O_2 delignification additives is evaluated experimentally, the experimental evaluation does not consider all the criteria at the same time while selecting the most appropriate additive (Bhardwaj et al., 2017). This chapter aims to present a WASPAS-based framework for the efficient selection of O_2 delignification additives.

Against this background, this chapter anticipates exploring the standard deviation-coupled WASPAS method for selecting suitable chelating agents, poly-meric additives, and chemical additives as cellulose protectors for the O_2 delignification process in the pulp and paper industry, based on the corresponding pulp and paper properties. For this purpose, three case studies have been formulated based on the literature data. The WASPAS method involves the determination of an optimal value of a term λ for a said decision-making situation. Therefore, this chapter also illustrates the effect of varying λ on the ranking of O_2 delignification additives. The remainder of the chapter is organized as follows: section 5.2 presents the research methodology adopted for this study, namely, data collection, operating procedure of WASPAS method, and calculation of objective weights by standard deviation method; section 5.3 discusses the beneficial and non-beneficial criteria and illustrates three case studies related to the selection of (a) chelating agents, (b) poly-meric additives, and (c) chelating agents and chemical additives to be used for O_2 delignification; and, ultimately, section 5.4 draws the pertinent conclusion from this study and highlights the limitations as well as scope for future research.

5.2 RESEARCH METHODOLOGY

5.2.1 DATA COLLECTION

Data for conducting this study was collected from a research report (Bhardwaj et al., 2017). The data comprised of different O_2 delignification additives i.e., chelating agents, polymeric additives, and chemical additives. The properties of O_2 delignified pulp such as yield, kappa number, intrinsic viscosity, and brightness drainage index; the properties of handsheets made from O_2 delignified pulp such as bulk, tensile index, burst index, tear index, porosity, double fold and smoothness; and the properties of pulp subsequently bleached using $DE_{OP}D$ (D = chlorine dioxide; E_{OP} = extraction with oxygen in presence of peroxide) sequences such as bright-ness, yield, and intrinsic viscosity are also included in the data. Three case studies of O_2 delignification additive selection were formulated based on the type of O_2 delignification additives used in the said report. The data so gathered was used to construct the decision matrices in the case studies presented later in the chapter.

5.2.2 OPERATING PROCEDURE OF THE WASPAS METHOD

Implementing WASPAS to a decision-making problem requires six steps. The first step aims at forming a decision matrix (DM). A DM contains alternatives, criteria, and performance values (pv) of all criteria associated with each alternative. A selection problem comprising of $A_i(i = 1,2,3,...,m)$ alternatives; $C_j(j = 1,2,3,...,n)$ criteria; and $pv_{ij}(i = 1,2,...,m; j = 1,2,...,n)$ performance values can be represented as DM in the following template:

	C_1	C_2	C_3	–	C_4
A_1	pv_{11}	pv_{12}	pv_{13}	–	pv_{1j}
A_2	pv_{21}	pv_{22}	pv_{23}	–	pv_{2j}
A_3	pv_{31}	pv_{32}	pv_{33}	–	pv_{3j}
–	–	–	–	–	–
A_i	pv_{i1}	pv_{i2}	pv_{i3}	–	pv_{ij}

The order of this DM can be expressed as $m \times n$. Here, m represents the number of alternatives and n the number of criteria. The second step requires the normalization of the decision matrix to produce normalized performance values (npv_{ij}). Normalization removes the heterogeneity of the decision matrix, makes the performance values dimensionless, and brings them to an equivalent scale. Normalization necessitates the identification of beneficial and non-beneficial criteria. @@@The details of beneficial and non-beneficial criteria are provided elsewhere in the chapter. Once the beneficial and non-beneficial criteria are recognized, normalization is carried out using the linear technique as shown below:

$$npv_{ij} = \begin{cases} \dfrac{pv_{ij}}{\max_i pv_{ij}} \text{for beneficial criteria} \\ \dfrac{\min_i pv_{ij}}{pv_{ij}} \text{for non} - \text{beneficial criteria} \end{cases} \quad (5.1)$$

The third step involves the calculation of the total relative significance of *ith* alternative as the first criterion of optimality based on the concept of WSM as indicated below:

$$Q_i^{(1)} = \sum_{j=1}^{n} npv_{ij} \times w_j \quad (5.2)$$

The fourth step entails the calculation of the total relative significance of *ith* alternative as the second criterion of optimality based on the concept of WPM as mentioned below:

$$Q_i^{(2)} = \prod_{j=1}^{n} npv_{ij}^{w_j} \quad (5.3)$$

The fifth step comprises the estimation of the combined criterion of optimality based on the WSM and WPM concept as presented above. The combined criterion of optimality can be estimated by the following relation:

$$Q_i = 0.5 \left(\sum_{j=1}^{n} npv_{ij} \times w_j + \prod_{j=1}^{n} npv_{ij}{}^{w_j} \right) \tag{5.4}$$

This can also be presented by the following generalized equation:

$$Q_i = \lambda \sum_{j=1}^{n} npv_{ij} \times w_j + (1 - \lambda) \prod_{j=1}^{n} npv_{ij}{}^{w_j}, \lambda = 0, 0.1, 0.2, 0.3 \ldots, 1 \tag{5.5}$$

Finally, in the sixth step, the alternatives are arranged in ascending order based on Q_i values. Thus, the most desired alternative gets the first position while the least desired alternative receives the last position.

5.2.3 STANDARD DEVIATION METHOD TO CALCULATE CRITERIA WEIGHT

The standard deviation procedure was employed to calculate the weights of several O_2 delignification criteria considered in this study. The weights obtained through this procedure were objective. The objective weights are free from the influence of preferences of a decision-maker. These depend solely on the characteristics of the decision matrix considered for a said problem. Determining the weights of criteria using standard deviation requires four steps. The first step encompasses the construction of a decision matrix (DM) for a given decision making problem. This DM is the same as constructed in the WASPAS method. The second step involves the normalization of the decision matrix for computing the normalized performance values (npv_{ij}). These normalized performance values were calculated using the following equation:

$$npv_{ij} = pv_{ij} / \sqrt{\left(\sum_{i=1}^{m} pv_{ij}^2 \right)} \tag{5.6}$$

The third step includes finding the standard deviation of the normalized performance values. The standard deviation (σ_j) of the normalized performance values was calculated using the following equation:

$$\sigma_j = \sqrt{\frac{\sum_{i=1}^{m} [npv_{ij} - \overline{npv_j}]^2}{m}}, i = 1, \ldots, m; j = 1, \ldots, n \tag{5.7}$$

The fourth step includes finding the weight of the criteria (w_j). For this purpose, the following equation is implemented:

$$w_j = \sigma_j / \sum_{j=1}^{n} \sigma_j \qquad\qquad (5.8)$$

The summation of all the weights of the criteria (w_j, $j = 1$, ...,n) of a decision-making problem is equal to one.

5.3 RESULTS AND DISCUSSION

5.3.1 IDENTIFICATION OF BENEFICIAL AND NON-BENEFICIAL CRITERIA

WASPAS analysis requires the identification of decision-making criteria as beneficial and non-beneficial. These are also referred to as benefit and cost criteria, respectively. A beneficial criterion is the one whose higher value is advantageous while a non-beneficial criterion is the one whose lower value is favorable. In WASPAS, these criteria come into play when linear normalization of the decision matrix is to be done. Among all the criteria considered in this study, the kappa number is a non-beneficial criterion while the remaining are beneficial criteria. From the papermaking perspective, the higher values of yield, viscosity, drainage index, brightness, bulk, tensile index, porosity, double-fold, smoothness, burst index, and tear index are desired whilst the lower values of kappa number are intended.

Kappa number denotes the residual lignin present in the unbleached pulp. It indicates bleachability or degree of delignification of the unbleached pulp. A lower kappa number is desired to achieve higher brightness, better bleachability, and lower bleaching chemical demand of the pulp (Hart & Connell, 2006). The yield of the pulp denotes the feasibility of commercial products related to a pulping process. The higher the yield, the more the desired pulp will be produced per ton of the raw material for papermaking. The improvement in pulp yield not only minimizes the raw material and bleaching cost but also maximizes the total solid loads to chemical recovery (Hart & Connell, 2006). The viscosity of a papermaking pulp refers to the degree of polymerization of the cellulose chains present in it. It provides an estimation for the intensity of deterioration of cellulose during pulping and bleaching actions. The lower the value of the viscosity the higher the degradation in the pulp and vice versa. The viscosity of the pulp affects the strength properties of paper sheets (Oglesby et al., 2017). It has been found that the strength properties of pulp depreciate strikingly when the degree of polymerization and hence the viscosity drops lower than a critical point sheet (Oglesby et al., 2017). Thus, a lower kappa number while maintaining a higher viscosity and adequate pulp yield is desired during optimization of a pulping process (McDonough et al., 1985).

The drainage index of a pulp, irrespective of raw material and method of processing, quantifies its drainability in water suspension and is an indirect measurement of its stiffness, compressibility, and partial density (Bonfiglio et al., 2013). The higher the value of the drainage index, the slower the draining. Also, with the increase in drainage index, there is an increase in charge ratio, drainage time, specific surface area, water retention value, tensile index, and burst index (Banavath et al., 2011). Further, a higher drainage index is achieved via refining of the pulp. The bulk, tensile index, double-fold, smoothness, burst index, and tear index are the mechanical strength properties of paper sheets. A higher value of these parameters is always desired. These properties in principle depend on the network, morphology, and structure of individual fibres present in the paper sheet. Eucalyptus fibres are stiff. This results in higher bulk. Higher bulk gives more open sheet structure and better strength during the wet state. This is the peculiarity of eucalyptus pulp over another hardwood. The combination of higher bulk and higher porosity not only improves drying performance at elevated machine velocity but also lowers the drying expense. Hence, the printability and runnability become better with higher bulk (Outreach, 2019).

5.3.2 Case 1: Selection Among Chelating Agents for O_2 Delignification

Bhardwaj et al. (2017) investigated the effect of five chelating agents on the O_2 delignification of a eucalyptus kraft pulp with initial kappa number 19.3, intrinsic viscosity 11.3 cP, and brightness 29.7%ISO. The five chelating agents used were HEDPA (hydroxyethylidenediphosphonic acid), NTA (nitrilotriacetic acid), EDTA (ethylenediaminetetraacetic acid), DTPA (diethylenetriaminepentaacetic acid), and DTMPA (diethylenetriamine pentamethylene phosphonic acid). Figure 5.1 shows the molecular structure of all of these chelating agents. The dose of these chelating agents maintained during O_2 delignification was 1%, 0.5%, 0.2%, 0.07% and 0.7% respectively. All O_2 delignification experimental trials were performed by maintaining pulp consistency 10%, temperature 95°C, time 60 min, O_2 1.8%, and NaOH 1.9%.

It was found that O_2 delignification of the eucalyptus kraft pulp without any chelating agent (hereafter referred to as Control – A) resulted in a 45.1% reduction in kappa number, 57.91% increase in brightness, and 17.70% reduction in intrinsic viscosity. However, with the addition of HEDPA, NTA, EDTA, DTPA and DTMPA, there were 46.6%, 47.2%, 46.1%, 49.2% and 47.7% reduction in kappa number; 15.93%, 12.39%, 4.42%, 15.04% and 17.70% reduction in intrinsic viscosity; while 58.25%, 61.95%, 49.49%, 65.99%, 60.94% increase in brightness, respectively. Thus, based on performance in terms of kappa number and brightness DTPA can be considered as the best while in terms of intrinsic viscosity EDTA can be rated as the best. Nevertheless, it is also important to note that the performance of EDTA in terms of brightness is worse even than that of the control. Generally, after O_2 delignification a significant reduction in kappa number, a sufficient increase in brightness, and a little decrease in intrinsic viscosity from that of the initial brown pulp are expected. Thus, from the aforementioned observations, it can be said that not a single chelating agent gives all the favourable and desired traits to the eucalyptus kraft pulp after O_2 delignification. Seeing this conflicting nature of O_2 delignification criteria, it was decided to couple the WASPAS multi-criteria decision-making method with the experimental data for effective selection of the best chelating agent based on a composite score that considers the contribution of all the criteria.

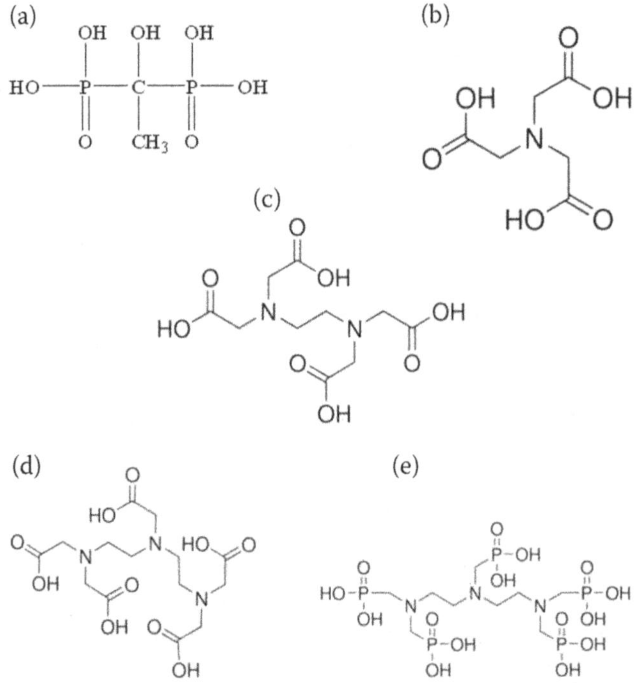

FIGURE 5.1 Structural and molecular formula of chelating agents used for O_2 delignification (a) HEDPA ($C_2H_8O_7P_2$), (b) NTA ($C_6H_9NO_6$), (c) EDTA ($C_{10}H_{16}N_2O_8$), (d) DTPA ($C_{14}H_{23}N_3O_{10}$) and (e) DTMPA ($C_9H_{28}N_3O_{15}P_5$).

Table 5.1 shows the decision matrix considered for this purpose. This table shows the kappa number, intrinsic viscosity, and brightness data of all the eucalyptus kraft pulp that were O_2 delignified with and without chelating agents. It is a decision matrix with six alternatives (Control-A, HEDPA, NTA, EDTA, DTPA, DTMPA) and three criteria (kappa number, brightness, intrinsic viscosity). Table 5.2 represents the normalized decision matrix obtained after implementing Eq. (5.1). Thus, for normalizing the kappa number data, the least value among the kappa number was divided by all the values of the kappa number, while for normalizing the brightness and intrinsic viscosity data, all the brightness and intrinsic viscosity values were divided by the highest value of brightness and intrinsic viscosity, respectively. This is so because kappa number is a non-beneficial criterion while brightness and intrinsic viscosity are beneficial criteria.

For further calculation of the $Q^{(1)}$, $Q^{(2)}$, and Q values, the weight of the O_2 delignification criteria needs to be incorporated. The weight of each of the O_2 delignification criteria was determined using the standard deviation method as described in Section 5.2.3. This calculation yielded the weightage of kappa number, brightness, and intrinsic viscosity as 22.31%, 29.08%, and 48.70% respectively making the aggregate ~100%. With the help of these weights, the $Q^{(1)}$, $Q^{(2)}$, and Q values were calculated as shown in Table 5.2 using Eq. (5.2), Eq. (5.3), and Eq.

TABLE 5.1

Decision Matrix for Selection of Chelating Agents for O_2 Delignification

Chelating agents	Kappa number	Brightness, %ISO	Intrinsic viscosity, cP
Control-A	10.6	46.9	9.3
HEDPA	10.3	47	9.5
NTA	10.2	48.1	9.9
EDTA	10.4	44.4	10.8
DTPA	9.8	49.3	9.6
DTMPA	10.1	47.8	9.3

TABLE 5.2

Normalized Decision Matrix and Q Values of Celating Agents

Chelating agents	Kappa number	Brightness, %ISO	Intrinsic viscosity, cP	$Q^{(1)}$	$Q^{(2)}$	Q
Control-A	0.9245	0.9513	0.8611	0.9023	0.9005	0.9014
HEDPA	0.9515	0.9533	0.8796	0.9179	0.9163	0.9171
NTA	0.9608	0.9757	0.9167	0.9445	0.9432	0.9439
EDTA	0.9423	0.9006	1.0000	0.9591	0.9572	0.9582
DTPA	1.0000	1.0000	0.8889	0.9468	0.9443	0.9455
DTMPA	0.9703	0.9696	0.8611	0.9178	0.9153	0.9165

(5.4) respectively. Based on the performance values obtained through the WASPAS method, the ranking order of chelating agents for O_2 delignification was received as EDTA – DTPA – NTA – HEDP – DTMPA – Control-A for a λ = 0.5. It has been found in the literature that the ranking of alternatives gets altered on altering the values of λ. Therefore, an attempt was made to evaluate the effect of changing λ on the ranking of chelating agents. Table 5.3 illustrates the effect of switching λ on the ranking order of chelating agents derived using the WASPAS method. It is stimulating to find that with the change in λ there is no change in the ranks of the chelating agents.

Figure 5.2 shows the variation of Q values of chelating agents concerning λ. It can be observed that at the higher values of λ, the Q values become better. This confirms that at higher values of λ WASPAS performs in the same way as WSM. Thus, EDTA can be selected as the best performer among all the available chelating agents to be used in O_2 delignification. It is also evident that the Control-A gets the last position. It ascertains that using chelating agents during O_2 delignification is advantageous to improve the pulp quality as well as to protect the cellulose degradation. Actually, the metal ions present in the pulp stimulates the formation of

TABLE 5.3

Effect of λ on the Ranking of Chelating Agents Using WASPAS Method

Chelating agents	$\lambda=0$	$\lambda=0.1$	$\lambda=0.2$	$\lambda=0.3$	$\lambda=0.4$	$\lambda=0.5$	$\lambda=0.6$	$\lambda=0.7$	$\lambda=0.8$	$\lambda=0.9$	$\lambda=1.0$
Control-A	0.9005(6)	0.9006(6)	0.9008(6)	0.9010(6)	0.9012(6)	0.9014(6)	0.9015(6)	0.9017(6)	0.9019(6)	0.9021(6)	0.9023(6)
HEDPA	0.9163(4)	0.9164(4)	0.9166(4)	0.9167(4)	0.9169(4)	0.9171(4)	0.9172(4)	0.9174(4)	0.9176(4)	0.9177(4)	0.9179(4)
NTA	0.9432(3)	0.9433(3)	0.9435(3)	0.9436(3)	0.9437(3)	0.9439(3)	0.9440(3)	0.9441(3)	0.9442(3)	0.9444(3)	0.9445(3)
EDTA	0.9572(1)	0.9574(1)	0.9576(1)	0.9578(1)	0.9580(1)	0.9582(1)	0.9584(1)	0.9586(1)	0.9587(1)	0.9589(1)	0.9591(1)
DTPA	0.9443(2)	0.9445(2)	0.9448(2)	0.9450(2)	0.9453(2)	0.9455(2)	0.9458(2)	0.9460(2)	0.9463(2)	0.9465(2)	0.9468(2)
DTMPA	0.9153(5)	0.9155(5)	0.9158(5)	0.9160(5)	0.9163(5)	0.9165(5)	0.9168(5)	0.9170(5)	0.9173(5)	0.9175(5)	0.9178(5)

Numbers in parentheses show the rank of chelating agents obtained on changing the values of λ.

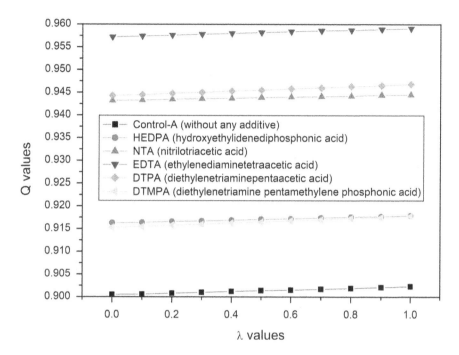

FIGURE 5.2   Variation of Q values of chelating agents concerning λ.

hydroxyl radicals responsible for damaging the pulp carbohydrates. The chelating agents confiscate these metal ions during O_2 delignification and thus hinders the carbohydrate degradation. Though the use of chelating agents maintains the intrinsic viscosity of the pulp during O_2 delignification, the extent of delignification gets reduced (Ragauskas et al., 2007).

5.3.3   CASE 2: SELECTION AMONG POLYMERIC ADDITIVES FOR O_2 DELIGNIFICATION

Bhardwaj et al. (2017) explored the efficacy of some polymeric additives towards the O_2 delignification. This was accomplished using eucalyptus kraft pulp having initial kappa number 19.7 and brightness 30%ISO. The polymeric additives used were native starch, cationic starch, CMC (carboxymethyl cellulose), and guar gum. The efficacy of O_2 delignification was evaluated in terms of the properties of O_2 delignified pulp and its paper handsheets. The pulp properties considered were yield, intrinsic viscosity, brightness, kappa number, and drainage index. The characteristics of paper handsheets evaluated were bulk, tensile index, burst index, tear index, porosity, double fold, and smoothness. In the above O_2 delignification trials, the dose of CMC was maintained at 0.1% while other polymeric additives were maintained at 0.5%. The pulp consistency, temperature, time, O_2, and NaOH maintained during entire O_2 delignification experiments were 10%, 95°C, 60 min, 1.8%, and 1.9%, respectively.

This study found that the eucalyptus kraft pulp on O_2 delignification without using any polymeric additive (hereafter referred to as Control-B) exhibited the lowest kappa number (44.16% reduction from that of initial kappa number), the highest brightness (52.67% increase from that of initial brightness) and the highest smoothness. However, this control pulp had a similar yield and drainage index, as well as its paper handsheets, which had also identical bulk to that of all the pulp O_2 delignified with polymeric additives. Nevertheless, the use of guar gum imparted the highest tensile index and burst index while the use of CMC resulted in the highest tear index, porosity, and double fold of the paper handsheets. It is interesting to note that the performance of O_2 delignification without any polymeric additive is even better in terms of kappa number, brightness, and smoothness than those with the polymeric additives. Thus, it can be said that not a single polymeric additive is sufficient enough to impart all the favourable properties to the O_2 delignified pulp. Hence, none can be adjudged as the best.

This problem of selecting the best polymeric additive for O_2 delignification can be attempted by the WASPAS multi-criteria decision-making approach. Table 5.4 shows the decision matrix constructed for this purpose. It contains the properties of the pulp and paper handsheets obtained after O_2 delignification with and without polymeric additives. Here, kappa number is the non-beneficial criteria while intrinsic viscosity, yield, drainage index, bulk, brightness, kappa number, tensile index, burst index, tear index, porosity, double fold, and smoothness are beneficial criteria. This decision matrix has five alternatives (Control-B, Cationic starch, Native starch, CMC, Guar gum) and twelve criteria (kappa number, brightness, intrinsic viscosity, yield, drainage index, bulk tensile index, burst index, tear index, porosity, double fold, smoothness).

Table 5.5 shows the normalized decision matrix. Normalization of kappa number data was performed by dividing its least value by all the values of kappa number because it is a non-beneficial attribute. On the other hand, the normalization of data of other attributes was done by dividing all the values by the highest value of that attribute because they are beneficial criteria. The main aim was to keep all the normalized values ≤1. This is what it is expressed in Eq. (5.1). The weights of these O_2 delignification criteria calculated using standard deviation method is as follows: kappa number = 0.0691; brightness = 0.0270; intrinsic viscosity = 0.0403; yield = 0.0031; °SR = 0.0099; bulk = 0.0109; tensile index = 0.0381; burst index = 0.0694; tear index = 0.1503; porosity = 0.1146; double fold = 0.3486; and smoothness = 0.1187. The summation of all the weights of these O_2 delignification criteria is one.

$Q^{(1)}$, $Q^{(2)}$, and Q values calculated using Eq. (5.2), Eq. (5.3), and Eq. (5.4) are shown in Table 5.6. Based on the Q values, the ranking order of polymeric additives can be derived as CMC – Cationic starch – Native starch – Guar gum – Control-B. Sometimes a change in the value of λ changes the ranking preference in WASPAS. Therefore, it is advised to perform the sensitivity analysis. Table 5.7 illustrates the effect of altering λ on the rank derived using WASPAS. It is obvious from this table that the ranking of polymeric additives remains unaltered with a change in λ. Figure 5.3 shows the variation of Q values of polymeric additives concerning λ. The larger the values of λ, the better the Q values. This substantiates that at larger values of λ WASPAS works similar to WSM. Hence, CMC can be considered as the best

TABLE 5.4
Decision Matrix for Selection Among Polymeric Additives for O_2 Delignification

Polymeric additives	Kappa number	Brightness, %ISO	Intrinsic Viscosity, cP	Yield, %	Drainage, °SR	Bulk, cm³/g	Tensile index, N.m/g	Burst index, kN/g	Tear index, mN.m²/g	Porosity, sec/ 100 ml	Double Fold	Smoothness, ml/min
Control-B	11	45.8	9	97	30	1.38	63.5	4.06	7.8	16.1	65.8	132
Cationic starch	11.6	44.4	9.6	97.5	30	1.4	63.9	4.4	8.6	19.1	139	120
Native starch	11.9	43.3	9.9	97.3	29.5	1.4	66.9	4.21	9.05	19	91.5	116
CMC	11.4	44.5	9.8	97.6	30	1.4	64.3	4.5	10.5	21.2	145	105
Guar gum	13	43	9.4	97.7	29.5	1.42	68.9	4.8	7.4	21	89	100

TABLE 5.5

Normalized Decision Matrix of Polymeric Additives

Polymeric additives	Kappa number	Brightness, %ISO	Intrinsic Viscosity, cP	Yield, %	Drainage, °SR	Bulk, cm³/g	Tensile index, N.m/g	Burst index, kN/g	Tear index, mN.m²/g	Porosity, sec/100 ml	Double Fold	Smoothness, ml/min
Control-B	1.0000	1.0000	0.9091	0.9928	1.0000	0.9718	0.9216	0.8458	0.7429	0.7594	0.4538	1.0000
Cationic starch	0.9483	0.9694	0.9697	0.9980	1.0000	0.9859	0.9274	0.9167	0.8190	0.9009	0.9586	0.9091
Native starch	0.9244	0.9454	1.0000	0.9959	0.9833	0.9859	0.9710	0.8771	0.8619	0.8962	0.6310	0.8788
CMC	0.9649	0.9716	0.9899	0.9990	1.0000	0.9859	0.9332	0.9375	1.0000	1.0000	1.0000	0.7955
Guar gum	0.8462	0.9389	0.9495	1.0000	0.9833	1.0000	1.0000	1.0000	0.7048	0.9906	0.6138	0.7576

TABLE 5.6
$Q^{(1)}$, $Q^{(2)}$ and Q Values of Polymeric Additives Using WASPAS Method

Polymeric additives	$Q^{(1)}$	$Q^{(2)}$	Q	Ranking
Control-B	0.7257	0.6904	0.7080	5
Cationic starch	0.9219	0.9205	0.9212	2
Native starch	0.8077	0.7959	0.8018	3
CMC	0.9651	0.9626	0.9639	1
Guar gum	0.7767	0.7616	0.7691	4

polymeric additive for O_2 delignification. Also, the very last position occupied by the control indicates that it is advantageous to use polymeric additives for better O_2 delignified pulp quality.

Literature suggests that the CMC available on the fiber surface itself gets degraded by the attack of hydroxyl radicals during O_2 delignification while protects the cellulose inside the fibres from this degradation (Kontturi et al., 2008). This leads to the increase in the intrinsic viscosity of the CMC added O_2 delignified pulp. Though the increase in the selectivity is non-significant, there is a substantial improvement in the tear and tensile strength after bleaching of CMC treated O_2 delignified pulp. In a study on O_2 delignification of a softwood pulp with polymeric additives, it has been found that guar bean based additive significantly improved the O_2 delignification selectivity as compared to starch and CMC. Such differences may be attributed to the differences in the adsorptive behavior of the polymer additives during O_2 delignification (Violette, 2003).

5.3.4 CASE 3: SELECTION AMONG CHELATING AND CHEMICAL ADDITIVES FOR O_2 DELIGNIFICATION

Bhardwaj et al. (2017) evaluated the efficiency of some chelating agents (EDTA and DTPA), chemical additive $Mg(OH)_2$ (magnesium hydroxide), and a mixture of chemical additive H_2O_2 (hydrogen peroxide) and chelating agent DTPA (i.e., H_2O_2 + DTPA) towards O_2 delignification. The O_2 delignification experiments were carried out by maintaining 10% pulp consistency, 95/100°C temperature, 30/90 min time, 4.5/5 kg/cm^2 O_2 pressure, 2% NaOH, 0.1% EDTA, 0.1% DTPA, 0.2% $Mg(OH)_2$ and 0.6%+0.05% H_2O_2 + DTPA. The performance of pulp O_2 delignified with these agents was compared with the pulp O_2 delignified without any additive (hereafter referred to as Control-C). The kappa number, brightness, intrinsic viscosity, and yield of all the O_2 delignified pulp were found in the range 9.5–11, 48–55%ISO, 10.2–12 cm^3/g, and 97–98% respectively. The lowest kappa number and the highest brightness and intrinsic viscosity were obtained for H_2O_2 + DTPA additive. The yield for $Mg(OH)_2$ and H_2O_2+DTPA was the same but the highest. These O_2 delignified pulps were further bleached using the $DE_{OP}D$

TABLE 5.7

Effect of λ on the Ranking of Polymeric Additives Using WASPAS Method

Polymeric additives	λ=0	λ=0.1	λ=0.2	λ=0.3	λ=0.4	λ=0.5	λ=0.6	λ=0.7	λ=0.8	λ=0.9	λ=1.0
Control-B	0.6904(5)	0.6939(5)	0.6974(5)	0.7010(5)	0.7045(5)	0.7080(5)	0.7116(5)	0.7151(5)	0.7186(5)	0.7222(5)	0.7257(5)
Cationic starch	0.9205(2)	0.9207(2)	0.9208(2)	0.9209(2)	0.9211(2)	0.9212(2)	0.9214(2)	0.9215(2)	0.9216(2)	0.9218(2)	0.9219(2)
Native starch	0.7959(3)	0.7971(3)	0.7983(3)	0.7995(3)	0.8006(3)	0.8018(3)	0.8030(3)	0.8042(3)	0.8053(3)	0.8065(3)	0.8077(3)
CMC	0.9626(1)	0.9629(1)	0.9631(1)	0.9634(1)	0.9636(1)	0.9639(1)	0.9641(1)	0.9643(1)	0.9646(1)	0.9648(1)	0.9651(1)
Guar gum	0.7616(4)	0.7631(4)	0.7646(4)	0.7661(4)	0.7676(4)	0.7691(4)	0.7706(4)	0.7722(4)	0.7737(4)	0.7752(4)	0.7767(4)

Numbers in parentheses show the rank of polymeric additives obtained on changing the values of λ

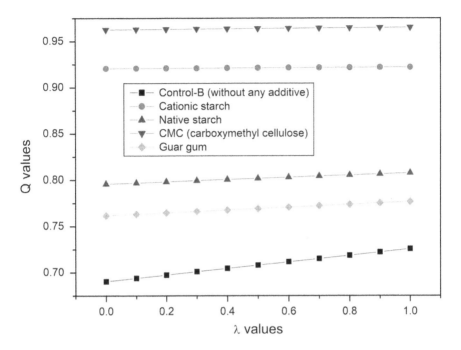

FIGURE 5.3   Variation of Q values of polymeric additives concerning λ.

bleaching sequence. The yield, brightness, and intrinsic viscosity of all the bleached pulps were in the range 41–44%, 87–89%ISO, and 9.4–10.5 cP, respectively. The highest brightness and the highest yield were obtained for DTPA + H_2O_2, the highest intrinsic viscosity was obtained for $Mg(OH)_2$, the lowest yield and intrinsic viscosity were obtained for EDTA. Thus, the conflicting nature of O_2 delignified pulps is quite evident. Therefore, none of them can be rated as the best. However, this decision making problem can be resolved by using the WASPAS multi-criteria decision-making method.

Table 5.8 illustrates the decision matrix established for this purpose. It contains the properties of the pulp obtained after O_2 delignification with and without chemical and chelating additives as well as after their subsequent $DE_{OP}D$ bleaching. This decision matrix has five alternatives (Control-C, EDTA, DTPA, Mg $(OH)_2$, DTPA+H_2O_2) and seven criteria (O_2D-Kappa number, O_2D-Brightness, O_2D-Viscosity, O_2D-Yield, $DE_{OP}D$-Brightness, $DE_{OP}D$-Yield, $DE_{OP}D$-Viscosity). In this decision matrix O_2D-Kappa number, O_2D-Brightness, O_2D-Viscosity, O_2D-Yield represent kappa number, brightness, intrinsic viscosity, and yield of pulp after O_2 delignification respectively while $DE_{OP}D$-Brightness, $DE_{OP}D$-Yield, and $DE_{OP}D$-Viscosity the brightness, yield, and intrinsic viscosity of pulp after $DE_{OP}D$ bleaching. Here, the kappa number is the only non-beneficial criteria while intrinsic viscosity, yield, and brightness are beneficial criteria. Table 5.9 demonstrates the normalized decision matrix. Normalization was performed using Eq. (5.1) accordingly for beneficial and non-beneficial criteria. In all the cases the normalized values are less than or equal to one.

TABLE 5.8

Decision Matrix for Selection Among Chelating and Chemical Additives for O$_2$ Delignification

Chelating and chemical additives	O$_2$D-Kappa number	O$_2$D-Brightness, %ISO	O$_2$D-Viscosity, cP	O$_2$D-Yield, %	DE$_{OP}$D-Brightness, %ISO	DE$_{OP}$D-Yield, %	DE$_{OP}$D-Viscosity, cP
Control-C	11	48	10.2	97	87	42	9.5
EDTA	10.8	49	11	97.5	88.2	41	9.4
DTPA	10.7	49.5	11.5	97.6	88.3	44	10
Mg(OH)$_2$	10	51	11.8	98	88.8	43.8	10.5
DTPA + H$_2$O$_2$	9.5	55	12	98	89	44	10.2

TABLE 5.9

Normalized Decision Matrix for Selection Among Chelating and Chemical Additives for O$_2$ Delignification

Chelating and chemical additives	O$_2$D-Kappa number	O$_2$D-Brightness, %ISO	O$_2$D-Viscosity, cP	O$_2$D-Yield, %	DE$_{OP}$D-Brightness, %ISO	DE$_{OP}$D-Yield, %	DE$_{OP}$D-Viscosity, cP
Control-C	0.8636	0.8727	0.8500	0.9898	0.9775	0.9545	0.9048
EDTA	0.8796	0.8909	0.9167	0.9949	0.9910	0.9318	0.8952
DTPA	0.8879	0.9000	0.9583	0.9959	0.9921	1.0000	0.9524
Mg (OH)$_2$	0.9500	0.9273	0.9833	1.0000	0.9978	0.9955	1.0000
DTPA + H$_2$O$_2$	1.0000	1.0000	1.0000	1.0000	1.0000	1.0000	0.9714

The weights of all the criteria considered in this case evaluated using standard deviation method are as follows: O$_2$D-Kappa number = 0.2227, O$_2$D-Brightness = 0.1999, O$_2$D-Viscosity = 0.2375, O$_2$D-Yield = 0.0157, DE$_{OP}$D-Brightness = 0.0326, DE$_{OP}$D-Yield = 0.1186 and DE$_{OP}$D-Viscosity = 0.1731. All these weights sum up to ~1. It can be seen that the O$_2$D-Viscosity is the most important criteria with 23.75% weightage followed by O$_2$D-Kappa number with 22.27% weightage, O$_2$D-Brightness with 19.99% weightage, and DE$_{OP}$D-Viscosity with 17.31%. Q$^{(1)}$, Q$^{(2)}$, and Q values calculated using Eq. (5.2), Eq. (5.3), and Eq. (5.4) are displayed in Table 5.10. Based on the Q values, the ranking order of additives can be established as DTPA+H$_2$O$_2$ – Mg(OH)$_2$ – DTPA – EDTA – Control-C.

Table 5.11 illustrates the result of sensitivity analysis performed to study the effect of altering λ on the rank derived using WASPAS. It is noticeable from this table that the ranking of additives remains intact with variation in λ. Figure 5.4

TABLE 5.10

$Q^{(1)}$, $Q^{(2)}$, Q Values and Ranking of Chelating and Chemical Additives Using WASPAS Method

Chelating and chemical additives	$Q^{(1)}$	$Q^{(2)}$	Q	Ranking
Control-C	0.8859	0.8850	0.8854	5
EDTA	0.9051	0.9046	0.9049	4
DTPA	0.9367	0.9357	0.9362	3
Mg(OH)$_2$	0.9699	0.9694	0.9696	2
DTPA + H$_2$O$_2$	0.9952	0.9950	0.9951	1

shows the variation of Q values of chelating agents and chemical additives concerning λ. It can be seen that with the increase in the values of λ the Q values increase. This reinforces that with increasing values of λ WASPAS resembles WSM. Hence, DTPA+H$_2$O$_2$ can be considered as the best additive for O$_2$ delignification. This ranking also suggests that instead of using chemical additives and chelating agents singly, their combination can be more favorable for O$_2$ delignification. Also, the very last position occupied by the Control-C indicates that it is advantageous to use additives for better O$_2$ delignification for subsequent bleaching though it exhibited better DE$_{OP}$D-Yield and DE$_{OP}$D-Viscosity than EDTA. Though Bhardwaj et al. (2017) also found DTPA+H$_2$O$_2$ the best O$_2$ delignification additive, they advocated the use of Mg(OH)$_2$ due to its lower cost.

5.4 CONCLUSION

This chapter illustrated the application of the WASPAS multi-criteria decision-making framework for the selection of suitable additives for O$_2$ delignification in the pulp and paper industry. The selection was done based on Q values obtained through the WASPAS method. Three cases dealing with O$_2$ delignification using (a) chelating additives (HEDPA, NTA, EDTA, DTPA, DTMPA); (b) polymeric additives (cationic starch, native starch, CMC, guar gum), and (c) chelating and chemical additives (EDTA, DTPA, Mg(OH)$_2$, DTPA+H$_2$O$_2$) were studied. In the first case, EDTA was found to be the best chelating agent; in the second case, CMC was identified as the best polymeric additive; while in the third case, DTPA+H$_2$O$_2$ emerged as the best additive. In all the cases, control O$_2$ delignification obtained the last position during ranking. This advocated the usefulness of using additives for O$_2$ delignification. In the first and third cases, the intrinsic viscosity was found to be the most influential criteria with 48.70% and 23.75% weightage, respectively. While in the second case, the double fold with 34.86% weightage was the most prominent. The weight of all the O$_2$ delignification criteria was objective and calculated using the standard deviation method. It was found in the literature that the experimental evaluation of these O$_2$ delignification additives does not consider all the criteria

TABLE 5.11

Effect of λ on the Ranking of Chelating and Chemical Additives Using WASPAS Method

Chelating and chemical additives	λ=0	λ=0.1	λ=0.2	λ=0.3	λ=0.4	λ=0.5	λ=0.6	λ=0.7	λ=0.8	λ=0.9	λ=1.0
Control-C	0.8850(5)	0.8851(5)	0.8852(5)	0.8852(5)	0.8853(5)	0.8854(5)	0.8855(5)	0.8856(5)	0.8857(5)	0.8858(5)	0.8859(5)
EDTA	0.9046(4)	0.9047(4)	0.9047(4)	0.9048(4)	0.9048(4)	0.9049(4)	0.9049(4)	0.9050(4)	0.9050(4)	0.9051(4)	0.9051(4)
DTPA	0.9357(3)	0.9358(3)	0.9359(3)	0.9360(3)	0.9361(3)	0.9362(3)	0.9363(3)	0.9364(3)	0.9365(3)	0.9366(3)	0.9367(3)
Mg (OH)$_2$	0.9694(2)	0.9694(2)	0.9695(2)	0.9695(2)	0.9696(2)	0.9696(2)	0.9697(2)	0.9697(2)	0.9698(2)	0.9698(2)	0.9699(2)
DTPA+H$_2$O$_2$	0.9950(1)	0.9950(1)	0.9950(1)	0.9950(1)	0.9951(1)	0.9951(1)	0.9951(1)	0.9951(1)	0.9951(1)	0.9951(1)	0.9952(1)

Numbers in brackets show the rank of chelating and chemical additives obtained on changing the values of λ

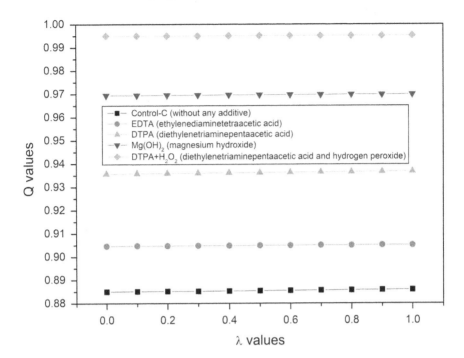

FIGURE 5.4 Variation of Q values of chelating agents and chemical additives concerning λ.

while selecting the most appropriate alternative. This study removed this anomaly. It was demonstrated that the experimental data when operated through a multi-criteria decision-making can lead to efficient decision making which is an important component in Industry 4.0. WASPAS proved to be an efficient decision-making tool for additive selection in the O_2 delignification process. No rank reversal problem was encountered with the change in λ. 0.5 can be assigned as the optimum λ value in all the cases. For further improving the decision-making and removing the limitations of this study, it is recommended to include cost, black liquor characteristics, effluent characteristics of these O_2 delignification additives, etc. in the decision matrix.

REFERENCES

Akhundzadeh, M., and Shirazi, B. (2017) Technology selection and evaluation in Iran's pulp and paper industry using 2-filterd fuzzy decision making method. *Journal of Cleaner Production*, 142, pp. 3028–3043. https://doi.org/10.1016/j.jclepro.2016.10.166

Anupam, K., Swaroop, V., Sharma, A.K., Lal, P.S., and Bist, V. (2015) Sustainable raw material selection for pulp and paper using SAW multiple criteria decision making design. *IPPTA Journal*, 27, pp. 67–76.

Anupam, K., Lal, P.S., Bist, V., Sharma, A.K., and Swaroop, V. (2014) Raw material selection for pulping and papermaking using TOPSIS multiple criteria decision making design. *Environmental Progress & Sustainable Energy*, 33(3), pp. 1034–1041. https://doi.org/10.1002/ep.11851

Anupam, K., Deepika, Swaroop, V., and Lal, P.S. (2018) Antagonistic, synergistic and interaction effects of process parameters during oxygen delignification of *Melia dubia* kraft pulp. *Journal of Cleaner Production*, 199, pp. 420–430. https://doi.org/10.1016/j.jclepro.2018.07.125

Banavath, H.N., Bhardwaj, N.K., and Ray, A.K. (2011) A comparative study of the effect of refining on charge of various pulps. *Bioresource Technology*, 102(6), pp. 4544–4551. https://doi.org/10.1016/j.biortech.2010.12.109

Bhardwaj, N.K., Tripathi, S.K., Marimuthu, P., Lal, P.S., and Sharma, A.K. (2017) Improvement in selectivity of oxygen bleaching. Report number ACIRD/CR/2K17-03. http://www.dcpulppaper.org/gifs/Report68.pdf

Bonfiglio, F., Curbelo, V., Santana, E., and Doldan, J. (2013) Influence of water conductivity on the drainability of Eucalyptus bleached Kraft pulp. In: 6th International Colloquium on Eucalyptus Pulp, Colonia del Sacramento, Uruguay, Nov. 24–27.

Brunnhofer, M., Gabriella, N., Schöggl, J.-P., Stern, T., and Posch, A. (2020) The biorefinery transition in the European pulp and paper industry – a three-phase Delphi study including a SWOT-AHP analysis. *Forest Policy and Economics*, 110, Art. no. 101882. https://doi.org/10.1016/j.forpol.2019.02.006

Chakraborty, S., and Zavadskas, E.K. (2014) Applications of WASPAS method in manufacturing decision making. *Informatica*, 25(1), pp. 1–20. https://doi.org/10.15388/informatica.2014.01

Chakraborty, S., Zavadskas, E.K., and Antucheviciene, J. (2015) Applications of WASPAS method as a multi-criteria decision-making tool. *Journal of Economic Computation and Economic Cybernetics Studies and Research*, 49(1), pp. 1–17.

Frias, A.R., da Silva, F.V., César, F.I.G., and Makiya, I.K. (2019) Industry 4.0: the importance of Automation in the Digital Transformation of the Pulp and Paper Mills in Brazil. *International Journal of Modern Research in Engineering and Technology*, 4(1), pp. 32–39.

Hart, P., and Connell, D. (2006) The effect of digester kappa number on the bleachability and yield of EMCC™ softwood pulp. *TAPPI Journal*, 5, pp. 23–27.

Kaushik, P., Tyagi, S., Thapliyal, B.P., and Anupam, K. (2015) An approach for selection of coating colour formulation using TOPSIS multi criteria decision making design. *TAPPSA Journal*, 6, pp. 20–25.

Kontturi, E., Mitikka-Eklund, Mia, and Vuorinen, T. (2008) Strength enhancement of fiber network by carboxymethyl cellulose during oxygen delignification of kraft pulp. *BioResources*, 3(1), pp. 34–45.

Kumar, R., Bhattacherjee, A., Singh, A.D., Singh, S., and Pruncu, C.I. (2020) Selection of portable hard disk drive based upon weighted aggregated sum product assessment method: a case of Indian market. *Measurement and Control*, 53(7–8), pp. 1218–1230. https://doi.org/10.1177/0020294020925841

Man, Y., Han, Y., Liu, Y., Lin, R., and Ren, J. (2020) Multi-criteria decision making for sustainability assessment of boxboard production: a life cycle perspective considering water consumption, energy consumption, GHG emissions, and internal costs. *Journal of Environmental Management*, 255, Art. no. 109860. https://doi.org/10.1016/j.jenvman.2019.109860

McDonough, T.J., VanDrunen, V.J., and Paulson, T.W. (1985) Sulphite-anthraquinone pulping of southern pine for bleachable grades. *Journal of Pulp and Paper Science*, 11, pp. 167–176.

Navarro, N., Daniela, P., Valverde, F., Quesada, H.J., and Madrigal-Sánchez, J. (2020) A supplier selection model for the wood fiber supply industry. *BioResources*, 15, pp. 1959–1977.

Oglesby, R.J., Moynihan, H.J., Santos, R.B., Ghosh, A., and Hart, P.W. (2017) Does kraft hardwood and softwood pulp viscosity correlate to paper properties? *TAPPI Journal*, 15, pp. 643–651.

Outreach (2019) *Farmer and planet friendly NewGen eucalyptus*. India: Outreach.

Ragauskas, A., Lucian, L., and Jameel, H. (2007) *High selectivity oxygen delignification*. Washington, DC: U.S. Department of Energy Office of Industrial Technologies. https://biorefinery.utk.edu/technical_reviews/High%20Selectivity%20Oxygen%20Delignification%20Final%20Report%20DOE.pdf

Stojčić, M., Zavadskas, E., Pamučar, D., Stević, Ž., and Mardani, A. (2019) Application of MCDM methods in sustainability engineering: a literature review 2008–2018. *Symmetry*, 11(3), Art. no. 350. https://doi.org/10.3390/sym11030350

Violette, S.M. (2003) Oxygen Delignification Kinetics and Selectivity Improvement, electronic theses and dissertations, p. 233. http://digitalcommons.library.umaine.edu/etd/233

Zavadskas, E.K., Turskis, Z., and Antucheviciene, J. (2012) Optimization of weighted aggregated sum product assessment. *Electronics and Electrical Engineering*, 122(6), pp. 3–6. https://doi.org/10.5755/j01.eee.122.6.1810

6 Does Green Consumption Matter? Insights from Emerging Markets

Shiv Kant Tiwari[1], Ashish Gupta[2], and Prashant Tiwari[3]

[1]Assistant Professor, Institute of Business Management, GLA University, Mathura (India)
[2]Assistant Professor, Department of Marketing, Indian Institute of Foreign Trade (IIFT), New Delhi (India)
[3]Assistant Professor, Institute of Business Management, GLA University, Mathura (India)

6.1 INTRODUCTION

India is a developing country and is witnessing rapid industrialization and infrastructure. But this rapid growth also has a harmful impact on the environment. Environmental consciousness is becoming a global phenomenon now. People across the world are becoming more aware of choosing products and services that are the result of environment-friendly business practices. Consumers across the world are looking at green, eco-friendly, recyclable, and natural products. A brand's environmental consciousness and eco-friendliness are rated as the highest parameters affecting shoppers, followed by foreign lifestyles, natural, and organic ingredients (Bhushan, 2019). In the last two decades, environmentalism has led to consumer acceptance of sustainable consumption (Kalafatis et al., 1999; Han et al., 2009). Customers are conscious of the environmental issues related to consumption. They are looking to buy environmentally acceptable products (Laroche et al., 2001; Kilbourne et al., 2009) for the benefit of future generations. However, satisfying personal needs continues to be a more important concern of consumers rather than protecting the environment (Verbeke et al., 2007; Dean et al., 2012). A key concern is sustainability: the balance between environment-friendliness, profit-generation, and people (Vermeir & Verbeke, 2008).

India is a favourable market for international players for reasons such as high-income class, over-population, high growth rate, and it is significant for the global market to get an idea of customer attitudes towards green consumption in the Indian context. Most researchers believe that a single customer's actions can be assumed

DOI: 10.1201/9781003102304-6

by their attitudes. In the past, research has attempted to improve the capacity to predict and understand customers' actions. These studies have also suggested several components, which can be categorized as situational or dispositional. Baker & Sinkula (2005) and Gupta & Tandon (2018) suggested that forecasting consumer behaviour is directly dependent on attitude. While measuring the attitude towards green marketing, we must measure the individual judgment towards the security and betterment of society. The demand for green products is based on customer attitude. Marketers have understood this and must put more effort to shift towards marketing green goods and services. Environment-friendly goods (Products) are referred to as eco-friendly products, ecological products, sustainable products, or green products (Ritter et al., 2015; Mont & Plepys, 2008). In recent years, industrial development has focussed more on the environment and sustainability. Different industry stakeholders need to be brought together to achieve sustainable consumption and output and to achieve the industrial sustainability target. Due to the various economic conditions and socio-cultural factors, sustainable consumption and growth require a broad focus in emerging and developed economies. There have been a few initiatives, in emerging and developed economies, to systematically compare the status and trend of sustainable consumption and growth. Indian firms are planning to change their working style and modify and diversify their activities for their survival in this competing environment (Chen, 2008).

The purpose of this study is to explore the factors leading to green consumption behaviour in emerging markets. Secondly, the study tries to identify the impact of green consumption practices on consumer attitude and purchase intention in the context of green products (Green Beverages in FMCG goods). Lastly, it suggests managerial practices to improve green consumption behaviour.

6.2 THEORETICAL FRAMEWORK AND HYPOTHESIS DEVELOPMENT

6.2.1 ENVIRONMENTAL CONCERN

Based on the extensive literature review selected key variables have been identified and a conceptual framework is also developed showing each relationship in Figure 6.1. Hu et al. (2010) addressed Environmental Concern ("EC") as the way in which people are conscious about environment-related difficulties and try to solve them or simply desire to personally contribute to the solution (Kim & Choi, 2005; Dunlap & Jones, 2002). Studies have examined the increasing attention of consumers toward environmental practices and the willingness to pay for sustainable products (Van Doorn & Verhoef, 2011). Hartmann and Apaolaza-Ibáñez (2012) examined the relationship of environmental concern with the attitude and purchase intention of consumers towards green brands. Customers have a preference for eco-friendly brands, and it is said that good and relevant messages in promotional strategies are more believable (Mathur & Mathur, 2000). EC is closely linked to the attitude towards green goods and weakly linked to green purchasing intentions (Khaola et al., 2014). Some researchers consider environmental concern as a significant indicator of customer attitude towards green products (Kirmani & Khan, 2016; Tang et al., 2014).

H1: *Environmental concern has a significant impact on attitude towards the purchase of green products.*

6.2.2 GREEN BRAND IMAGE

Chen (2010) defines green brand image as "a set of perceptions of a brand in a consumer's mind that is linked to environmental commitments and environmental concerns". The image of a green brand is found to affect the decision-making capacity and attitude of consumers (Jeong et al., 2014). The green brand image is the sum of a brand's characteristics and shortcomings that make an impact on the mind of the consumer. Brand image is a compilation of consumer emotions that imitate brand-by-brand affiliations (Cretu & Brodie, 2007). The green brand is not only a label in the present scenario, but it is much more than that which provides a base for the company to convey the information to the customers and improve market shares (Aaker, 1996). Green-brand image products are different from ordinary products in their performance and eco-friendliness. Currently, customers are influenced by those products and services that offer health benefits and have good quality (Karjaluoto & Chatterjee, 2009). In most markets around the world, there are products for which customers strongly demand green alternatives, such as plastic and pesticides, and customers highly value such alternatives (Rahbar & Wahid, 2011). More precisely, a positive connection between the green brand image and consumer attitude has been developed in the context of green goods (Khandelwal et al., 2019; Chen, 2010).

H2: *Green Brand Image has a significant impact on attitude towards the purchase of green products.*

6.2.3 GREEN PERCEIVED RISK

Green received risk is something that deals with the safety, security, and protection of the environment. Customers purchase a product based on the perceived risk, which affects their decision (Mitchell, 1999). Perceived risks have been broadly examined as an obstacle to the possibility of green purchases (Chen & Chang, 2013) and organic food (Gifford & Bernard, 2006). Nowadays, customers are very aware of the products and they get all the information, whatever the company promises, before making a purchase. Under such a condition, if companies try to cheat customers and do not fulfil their promise, the green perceived risk is increased (Boksberger et al., 2007). If something is very harmful to health and is promoted as a non-green product, it will affect the concept of green marketing (Ginsberg & Bloom, 2004). In previous studies, several forms of perceived risk have been described, including psychological aspects, physical aspects, product performance, social aspects, and convenience loss (Schiffman & Kanuk, 1994; Mitchell et al., 1999). Some authors have found an adverse relationship between green perceived risk and the purchase intention of consumers, as well as that the green perceived risk is a significant predictor of green purchase intention (Kim & Lennon, 2013; Chen & Chang, 2012).

H3: *Green Perceived Risk has a significant impact on attitude towards the purchase of green products.*

6.2.4 HEALTH CONSCIOUSNESS

In the last three decades, the consumption of unhealthy products/foods has become a big concern for consumers (Thomas & Mills, 2006). Health consciousness is described as "the degree to which health concerns are integrated into a person's daily activities, which is evaluated in terms of personal health-management characteristics" (Park et al., 2013, p. 332; Jayanti & Burns, 1998, p. 10). The important factor in the purchase of food items is health (Wandel & Bugge, 1997). Earlier studies have shown that consumers are more concerned about their health and safety while purchasing organic foods (Sirieix et al., 2011; Huber et al., 2011; Van Loo et al., 2010). Some studies have shown that health consciousness has a significant impact on customer attitude and purchasing intentions for organic food (Rana & Paul, 2017; Teng & Lu, 2016).

H4: *Health Consciousness has a significant impact on attitude towards the purchase of green products.*

6.2.5 CONSUMER ATTITUDE AND GREEN PURCHASE INTENTION

Attitude is characterized as a psychological path that decides an individual's favour or disfavour for a particular object (Eagly & Chaiken, 2007). Consumers prefer to indulge in action when they have a more positive attitude towards behaviour and plan to finish it (Ajzen, 1991). A positive attitude towards green goods dictates the intention of the customer to purchase green products (Vazifehdoust et al., 2013). In addition, recent research has shown that attitudes to organic food have a positive influence on purchasing intentions (Pino et al., 2012; Kim & Chung, 2011; Michaelidou & Hassan, 2008). Previous research on green products and environmental behaviour has also endorsed the argument that there is a positive relationship between attitude and green purchase intention (Albayrak et al., 2013; Yadav & Pathak, 2016; Barber et al., 2009; Teng, 2009; Mostafa, 2009; Aman et al., 2012; Chen, 2007; Sreen et al., 2018).

H5: *Attitudes has significant impact on purchase intentions towards the purchase of green products.*

6.3 RESEARCH METHODOLOGY

This study was causal in nature and a standardized questionnaire was adopted and modified as a tool for primary data collection. The scales of the study were adopted from studies (Table 6.1).

For data collection, the questionnaire was on the Five Point "Likert Type Scale", where 1 indicates the minimum agreement and 5 shows the maximum agreement.

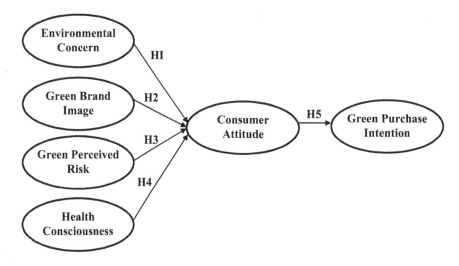

FIGURE 6.1   Conceptual framework.

This study was conducted in-person and online (Google docs); therefore, the sample frame was the consumers at Agra Region (U.P) and NCR (National Capital Region) of Northern India during the data collection phase of the study. A convenience sampling technique was used for data collection. In this study, we considered six variables and a sample size of 210 respondents was taken into consideration. Initially, about 290 questionnaires were distributed, of which 260 were returned. But after rejecting 50 questionnaires because they were incomplete, 210 were taken into consideration for further processing and analysis of the data.

To check the impact of the Independent variables on the Dependent variables, we used SPSS 20 as a statistical tool, and the Cronbach Alpha (α) Reliability Test was applied to check the reliability of the questionnaire. Factor analysis was applied to identify the underlying factors of all the studied variables i.e., Environmental Concern (EC), Green Brand Image (GBI), Green Perceived Risk (GPR), Health Consciousness (HC), Consumer Attitude (CATT), and Green Purchase Intention (GPI). Regression Analysis was applied to evaluate the impact of independent variables on the dependent variable.

6.4   RESULTS AND DISCUSSION

The authors conducted a reliability test using the Cronbach alpha coefficient, which shows the internal consistency of a multi-item scale for different structures (variables), used for data collection. Hair et al. (2006) suggested that the minimum value of coefficient alpha should be 0.6. The value of Cronbach's coefficient alpha was found to be within the threshold limit. For Environmental Concern, the value of alpha is 0.658. For Green Brand Image, Green Perceived Risk, Health Consciousness, the alpha value is 0.894, 0.867, 0.652, respectively, and these values were significant. The value of Cronbach's alpha for the studied variables (Consumer Attitude and Green Purchase Intention) comes out to be 0.838 and 0.882, which is considered significant (Table 6.1).

TABLE 6.1
Reliability and Validity Analysis

Construct	Item Code	Items	Item Loadings	Cronbach's Alpha
Environmental Concern (Kilbourne & Pickett, 2008)	EC_1	Environment is a major concern for me.	0.605	0.658
	EC_2	I will reduce my consumption for environmental protection.	0.721	
	EC_3	To protect the environment, political changes are required.	0.634	
	EC_4	To protect the environment, social changes are required.	0.512	
	EC_5	Anti-pollution rules should be strictly applied.	0.596	
Green Brand Image (Chen, 2010)	GBI_1	Brand should be dedicated for environmental commitments.	0.687	0.894
	GBI_2	The brand is professionally concern for environmental reputation.	0.706	
	GBI_3	The brand is effective in terms of environmental performance.	0.819	
	GBI_4	The brand is well established about environmental concern.	0.761	
	GBI_5	The brand possess credibility for environmental promises	0.749	
Green Perceived Risk (Jacoby & Kaplan, 1972; Murphy & Enis, 1986; and Sweeney et al., 1999)	GPR_1	There is a possibility that something will be negative with environmental performance of green beverages.	0.737	0.867
	GPR_2	There is a possibility that green beverages will not work smoothly with regard to its environmental design.	0.804	
	GPR_3	There is a possibility that you get environmental penalty or loss for using green beverages.	0.768	
	GPR_4	There is a possibility that using green beverages will negatively affect the environment.	0.796	
	GPR_5	Your reputation or image can be damaged if you use green beverages.	0.832	
Health Consciousness (Tarkiainen & Sundqvist,	HC_1	I believe that careful selection of green beverages will lead to good health.	0.528	0.652

TABLE 6.1 (Continued)
Reliability and Validity Analysis

Construct	Item Code	Items	Item Loadings	Cronbach's Alpha
2005; Petrescu et al., 2017)	HC_2	I believe that I am not a health-conscious consumer.	0.541	
	HC_3	I think about health-related issues regularly.	0.788	
	HC_4	I consider healthy living is important for me	0.511	
Consumer Attitude (McCarty & Shrum, 1994)	CATT_1	When making green beverages purchase, environmental protection is important to me.	0.652	0.838
	CATT_2	I believe that green beverages help in reducing pollution (water, air, etc.).	0.658	
	CATT_3	I believe that green beverages help in saving nature and its resources.	0.677	
	CATT_4	I will prefer green beverages over conventional beverages, if given a choice.	0.805	
Green Purchase Intention (Taylor & Todd, 1995; Ling-Yee, 1997; Mostafa, 2006, 2009; Chang & Chen, 2008)	GPI_1	As they are less polluting in the coming days, I will consider buying green beverages.	0.786	0.882
	GPI_2	For ecological reasons, I would consider switching to environment friendly brands.	0.71	
	GPI_3	Instead of traditional drinks, I plan to spend more on environment-friendly green beverages.	0.689	
	GPI_4	I expect to buy green beverages in the future as they are contributing to environment positively.	0.755	
	GPI_5	I want to buy green beverages in the near future.	0.696	

Total variance extracted: 65.742 %

6.5 TESTING OF HYPOTHESIS

We applied Regression Analysis in this study. We tested two models to achieve the objectives of the study. In model 1, we tested the relationship between Health consciousness, Green perceived risk, Environmental consciousness, and Green brand intention on Consumer attitude. The value of R=0.738 was found to be significant between consumer attitude and its predictors, which results in an R^2 value of 0.545. In Model 2, where the relationship between consumer attitude tested with Green purchase intention and relationship was found to be significant with a value of R= 0.835 and a resultant R^2 value of 0.697.

The Durbin-Watson test value in this model is around 2.197 (model 1) and 2.011 (model 2), respectively, which are within the criterion limit of 0 and 4 and shows that residuals are not correlated. The beta value explains the degree of each predictor influencing the relationship with a dependent variable. In model 1, the Green brand image is found to be the most significant predictor of the consumer attitude with the value of $\beta = 0.378$, then Health consciousness followed by Environmental consciousness with β values of 0.373 and 0.147, respectively. In model 2, consumer attitude was found to have a strong relationship with Green purchase intention with a β value of 0.835 (Table 6.2).

The formulated hypotheses were tested to empirically test the proposed research model. In our study, all proposed hypotheses are accepted except the one that shows a non-significant relationship between the Green perceived risk and Consumer Attitude (Table 6.3). The result of the hypotheses testing indicates the relationship amongst the variables proposed in the conceptual framework. The result suggests that Health consciousness, Green brand image, Environmental consciousness are found to be significant predictors of Consumer attitude, and Consumer attitude was found to be a strong predictor of Green purchase intention.

The reason behind the non-significant relationship could be an element of risk involved in adopting and using green products in regular consumption. However,

TABLE 6.2
Result of Regression Analysis – Model Summary

Independent Variables	Model-1 Dependent Variable Consumer Attitude (CATT)			Independent Variables	Model-2 Dependent Variable Green Purchase Intention (PI)		
	β value	t- Statistics	p-value		β value	t-Statistics	p-value
HC	.373	5.678	.000	CATT	0.835	19.188	.000
GPR	.044	.765	.445				
EC	.147	2.305	.023				
GBI	.378	5.919	.000				
R		0.738				0.835	
R-Square (R^2)		0.545				0.697	
Adjusted R^2		0.533				0.695	
F-value		46.953				368.172	
Durbin-Watson Statistics		2.197				2.011	

TABLE 6.3
Summary of Hypothesis Testing

Hypotheses	Description	Result
H1	Environmental Concern→ Consumer Attitude	Accepted
H2	Green Brand Image → Consumer Attitude	Accepted
H3	Green Perceived Risk→ Consumer Attitude	Rejected
H4	Health Consciousness → Consumer Attitude	Accepted
H5	Consumer Attitude → Green Purchase Intention	Accepted

there is a gradual increase in the usage of green products across different categories is increasing like beverages, fashion, packaging, etc. Another reason that affects the risk perception of consumer attitude toward green products is the availability of branded green products in the markets.

Other important parameters in the context of green product usage are growing interest in health and wellness among consumers in daily life. Consumers are also becoming aware of environmental protection and choosing the products accordingly.

6.7 CONCLUSION AND LIMITATIONS

The increasing attention of consumers on the environment, health, and other relevant factors is motivating marketers to offer tremendous opportunities in offering a wide variety of goods. However, consumers also have resistance towards several categories of such products. It is also observed that there is a low level of awareness about the use of green or organic products in Tier-III and -IV cities, where consumers do not want to change their regular lifestyle and consumption habits.

For Indian consumers, branded green products are a mark of their aspiration for a better life and quality consumption. However, low health consciousness, low levels of environmental awareness, and consumer resistance are major concerns for the marketers to create the markets for such products. This study focuses on few key variables, such as health consciousness, environmental consciousness, green perceived risk, and green brand image, which impact consumer attitude and lead to green purchase intention. The significance of health and environmental consciousness ha been widely studied in previous studies. In our research, Health Consciousness and Green brand image emerge as strong variables to predict consumer attitude towards green purchase intention. Green perceived risk is associated negatively with consumer attitude since the product category taken in the study was beverages, which include green tea, drinks, juices and others, where consumers do not find any significant impact of these on the environment as compared to other products like cars, air-condition, refrigerators, etc.

This study is limited in terms of its design and scope only to Tier-II and Tier-III cities. It means that the perspective of the study is very narrow, involving respondents from Agra Region (U.P) and NCR (North Central Region) of Northern India. The sample size is only 210 respondents. So it is suggested that a larger sample size be taken so that more appropriate results can be obtained. The findings of the study may highlight the consumer attitude towards green products in the selected geographical context that was considered for this present study, and which needs to be extended to yield the more generalized findings.

REFERENCES

Aaker, D.A. (1996). Measuring brand equity across products and markets. *California Management Review*, 38(3), pp. 102–120.

Ajzen, I. (1991). The theory of planned behavior. *Organizational Behavior And Human Decision Processes*, 50(2), pp. 179–211.

Albayrak, T., Aksoy, Ş., and Caber, M. (2013). The effect of environmental concern and scepticism on green purchase behaviour. *Marketing Intelligence & Planning*, 31, 27–39.

Aman, A.L., Harun, A., and Hussein, Z. (2012). The influence of environmental knowledge and concern on green purchase intention the role of attitude as a mediating variable. *British Journal of Arts and Social Sciences*, 7(2), pp. 145–167.

Baker, W.E., and Sinkula, J.M. (2005). Environmental marketing strategy and firm performance: Effects on new product performance and market share. *Journal of the Academy of Marketing Science*, 33(4), pp. 461–475.

Barber, N., Taylor, C., and Strick, S. (2009). Wine consumers' environmental knowledge and attitudes: Influence on willingness to purchase. *International Journal of Wine Research*, 1, pp. 59–72.

Bhushan, R. (2019). Call it natural, ethical or green to sell it in India. Retrieved 24 July 2020, from https://economictimes.indiatimes.com/industry/cons-products/fmcg/call-it-naturalethical-or-green-to-sell-it-in india/articleshow/71830503.cms?utm_source=contentofinterest&utm_medium=text&utm_campaign=cppst

Boksberger, P.E., Bieger, T., and Laesser, C. (2007). Multidimensional analysis of perceived risk in commercial air travel. *Journal of Air Transport Management*, 13(2), pp. 90–96.

Caruana, R. (2007). A sociological perspective of consumption morality. *Journal of Consumer Behaviour: An International Research Review*, 6(5), pp. 287–304.

Chang, H.H., and Chen, S.W. (2008). The impact of online store environment cues on purchase intention. *Online Information Review*, 32, pp. 818–841.

Chen, M.F. (2007). Consumer attitudes and purchase intentions in relation to organic foods in Taiwan: Moderating effects of food-related personality traits. *Food Quality and Preference*, 18(7), pp. 1008–1021.

Chen, Y.S. (2008). The driver of green innovation and green image–green core competence. *Journal of Business Ethics*, 81(3), pp. 531–543.

Chen, Y.S. (2010). The drivers of green brand equity: Green brand image, green satisfaction, and green trust. *Journal of Business Ethics*, 93(2), pp. 307–319.

Chen, Y.-S., and Chang, C.H. (2012). Enhance green purchase intentions: The role of green perceived value, green perceived risk, and green trust. *Management Decision*, 50(3), pp. 502–520.

Chen, Y.S., and Chang, C.H. (2013), Towards green trust. *Management Decision*, 51(1), pp. 63–82.

Cretu, A.E., and Brodie, R.J. (2007). The influence of brand image and company reputation where manufacturers market to small firms: A customer value perspective. *Industrial Marketing Management*, 36(2), pp. 230–240.

Dean, M., Raats, M.M., and Shepherd, R. (2012). The role of self-identity, past behavior, and their interaction in predicting intention to purchase fresh and processed organic food. *Journal of Applied Social Psychology*, 42(3), pp. 669–688.

Dunlap, R., and Jones, R. (2002). Environmental concern: Conceptual and measurement issues. In *Handbook of Environmental Sociology*, ed. R. Dunlap and W. Michelson. London: Greenwood.

Eagly, A.H., and Chaiken, S. (2007). The advantages of an inclusive definition of attitude. *Social Cognition*, 25(5), pp. 582–602.

Gifford, K., and Bernard, J.C. (2006). Influencing consumer purchase likelihood of organic food. *International Journal of Consumer Studies*, 30(2), pp. 155–163.

Ginsberg, J.M., and Bloom, P.N. (2004). Choosing the right green marketing strategy. *MIT Sloan Management Review*, 46(1), pp. 79–84.

Gupta, A., and Tandon, A. (2018). Branding for bottom of the pyramid: A case of branded footwear consumer in Indian rural setting. In: Dwivedi, Y. et al. eds. *Emerging markets from a multidisciplinary perspective. Advances in theory and practice of emerging markets* . Cham: Springer.

Hair, J.F., Black, W.C., Babin, B.J., Anderson, R.E., and Tatham, R. (2006). *Multivariate Data Analysis*. Upper Saddle River.

Han, H., Hsu, L.T.J., and Lee, J.S. (2009). Empirical investigation of the roles of attitudes toward green behaviors, overall image, gender, and age in hotel customers' eco-friendly decision-making process. *International Journal of Hospitality Management*, 28(4), pp. 519–528.

Hartmann, P., and Apaolaza-Ibáñez, V. (2012). Consumer attitude and purchase intention toward green energy brands: The roles of psychological benefits and environmental concern. *Journal of Business Research*, 65(9), pp. 1254–1263.

Hu, H.H., Parsa, H.G., and Self, J. (2010). The dynamics of green restaurant patronage. *Cornell Hospitality Quarterly*, 51(3), pp. 344–362.

Huber, M., Rembiałkowska, E., Średnicka, D., Bügel, S., and Van De Vijver, L.P.L. (2011). Organic food and impact on human health: Assessing the status quo and prospects of research. *NJAS-Wageningen Journal of Life Sciences*, 58(3–4), pp. 103–109.

Jacoby, J., and Kaplan, L.B. (1972). The components of perceived risk Proceedings of the Annual Conference of the Association for Consumer Research. *ACR Special Volumes*, 10, pp. 382–393.

Jayanti, R.K., and Burns, A.C. (1998). The antecedents of preventive health care behavior: An empirical study. *Journal of the Academy of Marketing Science*, 26(1), pp. 6–15.

Jeong, E., Jang, S.S., Day, J., and Ha, S. (2014). The impact of eco-friendly practices on green image and customer attitudes: An investigation in a café setting. *International Journal of Hospitality Management*, 41, pp. 10–20.

Kalafatis, S.P., Pollard, M., East, R., and Tsogas, M.H. (1999). Green marketing and Ajzen's theory of planned behaviour: A cross-market examination. *Journal of Consumer Marketing*, 16(5), pp. 441–460.

Karjaluoto, H., and Chatterjee, P. (2009). Green brand extension strategy and online communities. *Journal of Systems and Information Technology*, 11(4), pp. 367–384.

Khandelwal, U., Kulshreshtha, K., and Tripathi, V. (2019). Importance of Consumer-based green brand equity: Empirical evidence. *Paradigm*, 23(1), pp. 83–97.

Khaola, P.P., Potiane, B., and Mokhethi, M. (2014). Environmental concern, attitude towards green products and green purchase intentions of consumers in Lesotho. *Ethiopian Journal of Environmental Studies and Management*, 7(4), pp. 361–370.

Kilbourne, W.E., Dorsch, M.J., Mc Donagh, P., Urien, B., Prothero, A., Grünhagen, M., and Bradshaw, A. (2009). The institutional foundations of materialism in western societies: A conceptualization and empirical test. *Journal of Macro Marketing*, 29(3), pp. 259–278.

Kilbourne, W., and Pickett, G. (2008). How materialism affects environmental beliefs, concern, and environmentally responsible behavior. *Journal of Business Research*, 61(9), pp. 885–893.

Kim, H.Y., and Chung, J.E. (2011). Consumer purchase intention for organic personal care products. *Journal of Consumer Marketing*, 28(1), pp. 40–47.

Kim, Y., and Choi, S.M. (2005). Antecedents of green purchase behavior: An examination of collectivism, environmental concern, and PCE. *ACR North American Advances*, 32(1), pp. 592–599.

Kim, J., and Lennon, S.J. (2013). Effects of reputation and website quality on online consumers' emotion, perceived risk, and purchase intention. *Journal of Research in Interactive Marketing*, 7, 33–56.

Kirmani, M.D., and Khan, M.N. (2016). Environmental concern to attitude towards green products: evidences from India. *Serbian Journal of Management*, 11(2), pp. 159–179.

Laroche, M., Bergeron, J., and Barbaro-Forleo, G. (2001). Targeting consumers who are willing to pay more for environmentally friendly products. *Journal of Consumer Marketing*, 18(6), pp. 503–520.

Ling-Yee, L. (1997). Effect of collectivist orientation and ecological attitude on actual environmental commitment: The moderating role of consumer demographics and product involvement. *Journal of International Consumer Marketing*, 9(4), pp. 31–53.

Mathur, L.K., and Mathur, I. (2000). An analysis of the wealth effects of green marketing strategies. *Journal of Business Research*, 50(2), pp. 193–200.

McCarty, J.A., and Shrum, L.J. (1994). The recycling of solid wastes: Personal values, value orientations, and attitudes about recycling as antecedents of recycling behavior. *Journal of Business Research*, 30(1), pp. 53–62.

Michaelidou, N., and Hassan, L.M. (2008). The role of health consciousness, food safety concern and ethical identity on attitudes and intentions towards organic food. *International Journal of Consumer Studies*, 32(2), pp. 163–170.

Mitchell, V.W. (1999). Consumer perceived risk: Conceptualisations and models. *European Journal of Marketing*, 33, 163–195.

Mitchell, V.W., Davies, F., Moutinho, L., and Vassos, V. (1999). Using neural networks to understand service risk in the holiday product. *Journal of Business Research*, 46(2), pp. 167–180.

Mont, O., and Plepys, A. (2008). Sustainable consumption progress: should we be proud or alarmed? *Journal of Cleaner Production*, 16(4), pp. 531–537.

Mostafa, M.M. (2006). Antecedents of Egyptian consumers' green purchase intentions: A hierarchical multivariate regression model. *Journal of International Consumer Marketing*, 19(2), pp. 97–126.

Mostafa, M.M. (2009). Shades of green: A psychographic segmentation of the green consumer in Kuwait using self-organizing maps. *Expert Systems with Applications*, 36(8), pp. 11030–11038.

Murphy, P.E., and Enis, B.M. (1986). Classifying products strategically. *Journal of Marketing*, 50(3), pp. 24–42.

Park, S.H., Yoon, H.J., Cho, S.H., and Haugtvedt, C.P. (2013). Assessing the provision of nutritional information on quick service restaurant menu item choices for college students. *Journal of Foodservice Business Research*, 16(4), pp. 329–346.

Petrescu, A.G., Oncioiu, I., and Petrescu, M. (2017). Perception of organic food consumption in Romania. *Foods*, 6(6), Art. no. 42.

Pino, G., Peluso, A.M., and Guido, G. (2012). Determinants of regular and occasional consumers' intentions to buy organic food. *Journal of Consumer Affairs*, 46(1), pp. 157–169.

Rahbar, E., and Wahid, N.A. (2011). Investigation of green marketing tools' effect on consumers' purchase behavior. *Business Strategy Series*, 12(2), pp. 73–83.

Rana, J., and Paul, J. (2017). Consumer behavior and purchase intention for organic food: A review and research agenda. *Journal of Retailing and Consumer Services*, 38, pp. 157–165.

Ritter, Á.M., Borchardt, M., Vaccaro, G.L., Pereira, G.M., and Almeida, F. (2015). Motivations for promoting the consumption of green products in an emerging country: exploring attitudes of Brazilian consumers. *Journal of Cleaner Production*, 106, pp. 507–520.

Schiffman, L.G., and Kanuk, L.L. (1994). *Consumer Behavior*. Englewood Cliffs, New Jersey, USA: Prentice Hall.

Sirieix, L., Kledal, P.R., and Sulitang, T. (2011). Organic food consumers' trade-offs between local or imported, conventional or organic products: A qualitative study in Shanghai. *International Journal of Consumer Studies*, 35(6), pp. 670–678.

Sreen, N., Purbey, S., and Sadarangani, P. (2018). Impact of culture, behavior, and gender on green purchase intention. *Journal of Retailing and Consumer Services*, 41, pp. 177–189.

Sweeney, J.C., Soutar, G.N., and Johnson, L.W. (1999). The role of perceived risk in the quality-value relationship: A study in a retail environment. *Journal of retailing*, 75(1), pp. 77–105.

Tang, Y., Wang, X., and Lu, P. (2014). Chinese consumer attitude and purchase intent towards green products. *Asia-Pacific Journal of Business Administration*, 6(2), pp. 84–96.

Tarkiainen, A., and Sundqvist, S. (2005). Subjective norms, attitudes, and intentions of Finnish consumers in buying organic food. *British Food Journal*, 107(11), pp. 808–822.

Taylor, S., and Todd, P. (1995). An integrated model of waste management behavior: A test of household recycling and composting intentions. *Environment and Behavior*, 27(5), pp. 603–630.

Teng, C.C., and Lu, C.H. (2016). Organic food consumption in Taiwan: Motives, involvement, and purchase intention under the moderating role of uncertainty. *Appetite*, 105, pp. 95–105.

Teng, L. (2009). A comparison of two types of price discounts in shifting consumers' attitudes and purchase intentions. *Journal of Business Research*, 62(1), pp. 14–21.

Thomas Jr, L., and Mills, J.E. (2006). Consumer knowledge and expectations of restaurant menus and their governing legislation: A qualitative assessment. *Journal of Foodservice*, 17(1), pp. 6–22.

Trudel, R., and Cotte, J. (2008). Does being ethical pay. *Wall Street Journal*, 1. https://www.wsj.com/articles/SB121018735490274425#:%7E:text=As%20in%20the%20first%20t est,production%20with%20increasing%20price%20premiums.

Van Doorn, J., and Verhoef, P. C. (2011). Willingness to pay for organic products: Differences between virtue and vice foods. *International Journal of Research in Marketing*, 28(3), pp. 167–180.

Van Loo, E., Caputo, V., Nayga, Jr, R.M., Meullenet, J.F., Crandall, P.G., and Ricke, S.C. (2010). Effect of organic poultry purchase frequency on consumer attitudes toward organic poultry meat. *Journal of Food Science*, 75(7), pp. S384–S397.

Vazifehdoust, H., Taleghani, M., Esmaeilpour, F., & Nazari, K. (2013). Purchasing green to become greener: Factors influence consumers' green purchasing behavior. *Management Science Letters*, 3(9), pp. 2489–2500.

Verbeke, W., Sioen, I., Brunsø, K., De Henauw, S., and Van Camp, J. (2007). Consumer perception versus scientific evidence of farmed and wild fish: exploratory insights from Belgium. *Aquaculture International*, 15(2), pp. 121–136.

Vermeir, I., and Verbeke, W. (2008). Sustainable food consumption among young adults in Belgium: Theory of planned behaviour and the role of confidence and values. *Ecological Economics*, 64(3), pp. 542–553.

Wandel, M., and Bugge, A. (1997). Environmental concern in consumer evaluation of food quality. *Food Quality and Preference*, 8(1), pp. 19–26.

Yadav, R., and Pathak, G.S. (2016). Young consumers' intention towards buying green products in a developing nation: Extending the theory of planned behavior. *Journal of Cleaner Production*, 135, pp. 732–739.

7 Energy Consumption Optimization Based on Six-Sigma Tool (DAMIC) and Energy Value Stream Mapping

Neha Verma[1] and Vinay Sharma[2]
[1]Shri Shankaracharya Institute of Professional Management and Technology (SSIPMT)
[2]Birla Institute of Technology Mesra, Ranchi

7.1 INTRODUCTION

Due to a higher degree of digitalization, which results in an informed, linked, and decentralized development, industrial production processes have been transformed in recent years. This latest stage is also referred to as "The Fourth Industrial Revolution" (Helbig 2013; Hermann et al., 2016). Industry 4.0's central concept is to use new technology to make manufacturing operations of industry and technological processes fundamentally integrated into a scalable, productive, and sustainable approach to high quality and low cost (Wang et al., 2016). In India, manufacturing industries are one of the top consumers of energies and contribute to one-third of the total energy consumption of the nation. The industry accounted for the biggest share of the overall electricity consumption of 44.11%, and for total electricity consumption in 2015–16, according to data obtained before 2014–2015. This represents the biggest share of the industry's consumption of 42.30% (Behera, 2015). Consequently, the issue of resource scarcity arises, as it is unclear whether economic development can be sustained in a world of natural limited resources (Krautkraemer 2005). The declining natural resource volume can be a concern for global supply chain production companies. For example, these businesses need to cope with increased resource prices and uncertainty about supply (Preston & Herron 2016). Compared to other sectors in 2006–07 to 2015–16 electricity consumption in the industry grew at an accelerated rate. The compounded annual growth rate was 9.47% and 7.97%. The electricity usage of factories is shown in Figure 7.1. This year is the abscissum axis and in giga watt-hour the axis reflects electricity consumed.

DOI: 10.1201/9781003102304-7

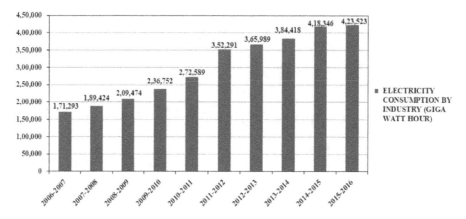

FIGURE 7.1   Electricity consumption by industry in India (Behera, 2015).

With Lean and Six-Sigma methods (Sihag et al., 2014), this rise in energy use can be tackled (Lee et al., 2014 and Baswaraj et al., 2015). Lean and Six Sigma are two principles that share the same methodologies and the same instruments, but in many ways they are distinct (Verma and Sharma 2015). Six Sigma focuses on waste management and cost savings (Kane and Dupont 2003). Six Sigma improves the reliability of process output by finding and eliminating the causes of defects and reducing production process variability (Kaushik et al., 2008). Lean management is focused on removing defects and reducing uncertainty (Kaushik and Khanduja, 2007). It primarily uses analytical, statistical, and quality control methods (Kumar and Antony 2008) and built a specific ecosystem of people with expertise in those methods within the organisation (Soni et al., 2013).

But, using capital, raw materials, information and energy with Industry 4. remains high and, despite several industry 4.0 advantages, environmentally unsustainable, which has increased the perception of society and the public sector of the environmental and risk issues (Müller et al., 2018). This paper therefore aims to change the conditions in this sector and to help it become environmentally and socially sustainable not just economically better. This sector with the use of Industry 4.0 focussed mainly on production and profit making, creating many other dimensions difficulties (Bettoni et al., 2015). For example, diminishing natural resources, adverse environmental impacts, income unbalance and poor working conditions can all lead to an environmentally, economically and socially unsustainable consumption pattern (Swain et al., 2018).

This study's main objective was to understand the extent to which Industrie 4.0 could, in the last review, have a negative or positive impact on environmental sustainability, affect the energy and resources floating raw materials as well as the waste and residues generated as a consequence of industry 4.0 technologies (Bunse and Vodicka 2011). For purposes, data have been analysed and each form summarised. The implications of Industry 4.0 on operating scenarios for companies were further analysed subsequently. In addition, their patterns have been investigated with negative and positive effects on the sustainability (Lin et al., 2017).

Today, the concept of Industry 4.0 is ubiquitous. Industry 4.0 is closely related to megatrends such as connectivity and digitalisation. The output in Industry 4.0 is related to the latest contact and IT (Federal Ministry of Economics and Energy 2016). In addition to all competitive advantages, such as productivity and versatility improvements, efficient utilisation of resources should also be considered (Bonilla et al., 2018). In this relation of the manufacturing processes in industry 4.0 holistic equilibrium circuits are expected to direct and shape the latest industrial development (ArbeitskreisIndustrie 4.0 2012, pp. 30–31). The burden on industrial enterprises has increased over time, due to environmental pollution (Rickand and Dugan 2008) and diminishing capital. These circumstances face manufacturing industries in response to pressure from government environmental legislation, resource price volatility challenges due to scarce resources and resource provision threats. A circle economy may be a solution for harmonising the objectives of economic development and protection of environment where the circular economy is seen in the closed loop material flow to be done in the economic system (Lieder & Rashid, 2016, pp. 36–51). The creation of Industry 4.0 provides tremendous opportunities to achieve sustainable manufacturing that is environmentally friendly and resource saving (Stock & Seliger 2016, p. 536–541).

A well defined sequence of steps follows each Six Sigma project undertaken in an organisation, which has a clear value goal. To minimise time, waste, expenditures and client satisfaction as well as profitability, for example.

7.2   CASE STUDY

The study was carried out at **Kalpataru Power Transmission Limited (KPTL).** There are two production facilities in Gandhinagar (Gujarat) and the third in Raipur (Chattisgarh). The manufacturing site in Raipur is located 24 km from the town centre of **Kalpataru Power Transmission Limited**. The business was founded in 2012. It is a big industry with 6000 Crores turnover. Hook, cleat and plates are the items that are made. In Figure 7.2, the factory design is shown.

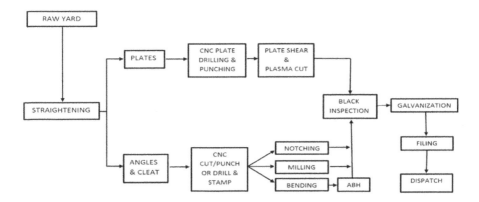

FIGURE 7.2   Raipur plant layout of Kalpataru.

FIGURE 7.3   Step of galvanizing operation.

7.2.1 Process Layout

There are two process layouts, one for manufacturing of plate and other to manu-facture angles and cleats.

In KPTL, 15 CNC machines are available. Bending is used to build a ductile material, most often sheet metal, V-shape, U-shape, or channel shape in a right axis.

The following measures in Figure 7.3 included the Galvanizing process.

7.3   INDUSTRY 4.0

The concept of industry 4.0 has been widely used in Europe, and in Germany in particular. Some commentators also use the words "internet of things," "internet of things" or "industrial internet" in the USA and the Englistically speaking world. The fact that traditional development and manufacturing processes are in the throes of digital transformation is similar to all these terms and meanings.

The concept 'Industry 4.0' derived from the German term 'Industry 4.0' was first mentioned publicly once during Germany's 2011 Hannover Trade Fair (Kagermann et al., 2016). The German Federal Minister of Education and Research (BBMF) – known as "Industry 4.0" (BMBF) – created an initiative aimed at encouraging German manufacturing to prepare a future for production.

The word Industry 4.0 became a popular concept in 2011 and in the German Government's initiative it was specifically called as an industrial growth strategy. This vision shows that Industry 4.0 forms part of an integral integrated world created by the ICT revolution. This technology transition happens in Industry 4.0 through the Internet of Things and the IOS, which links the industry internally and externally through the electronic supply chain network (Liao et al., 2017). The CPS discussed in the text can be explain by defining structures linked to large-scale hardware and software components to generate outcomes. The CPS can be described in the text. An review of literature of Industry 4.0's structural approach and its barriers to energy conservation (Da Silva et al., 2020). Industry 4.0 is also an important component of CPS and IoT. Many internet-related conceptsIndustry 4.0 such as IOS, Internet of Everything (IOE) and IOP have been demonstrated in previous studies. Industry 4.0, including big data, cloud computing, 3D pressing, and blockchain, contains several IOE and CPS technologies (Fu et al., 2018).

Intelligent machines share information on current levels of inventory, faults or fault and continuous order or demand changes. Processes and time limits are structured to enhance efficiency and to optimise throughput, use of growth skills and quality, development, marketing and purchases.

In addition, the CPPSs do not just link network machines but construct an intelligent network of machines, properties, ICT systems, intelligent products and individuals along the value chain and the entire product life cycle. Maschines can be associated with plants, fleets, networks and humans with sensors and control elements.

Intelligent networks such as these are the cornerstone of intelligent plants that are the foundation of industry 4.0 (see Figure 7.4).

In some manufacturing enterprises and business industries, this movement is still in its infancy, but in others transformation into Industry 4.0 is well underway.

7.3.1 SMART MANUFACTURING IN THE SMART FACTORY

In line with the vision of Industry 4.0, the future of production could look the following: the whole vertical value chain (creation of goods, production, services) is distributed through an array of apps and networks. Intelligent machines exchange real-time information and instructions around the supply chain and the entire product life cycle (PLC) with intelligent materials and people (Murugananthan et al., 2014). Plants, floats, networks and people can be linked by sensors and elements of power. Machinery itself exchanges knowledge about existing inventories, complaints, faults and orders or demand changes continuously (Timothy et al., 2011). Processes and timeframes are also structured in order to enhance efficiency and increase throughput times (Bertoldi and Mosconi 2020). This will boost performance in the PLC. This will build an autonomous and optimised production framework in its entirety (Siemens, 2014).

While Industry 4.0 focuses mainly on the manufacturing industry, concepts like Internet of Things, Digital Revolution and the Internet of All concentrate more on promoting and expediting the adoption of internet technology in manufacturing and

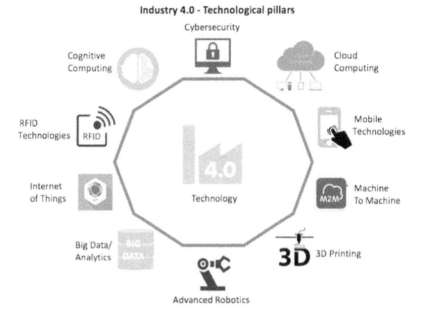

FIGURE 7.4   Industry–technological pillars.

non-manufacturing industries (Zawadzki and Zywicki 2016). Nevertheless, it is understood that traditional structures undergo a digital transformation that has all these terms and concepts in common (Deloitte, 2015).

7.3.2   CONCEPTS OF INDUSTRY 4.0 AND ITS TECHNOLOGIES

In 2011, the concept of Industry 4.0 was improved by a government initiative from Germany, established as an acritical plan for industrial growth. In this view Industry 4.0 has been transformed by the ICT revolution as part of the interconnecting environment. The IoT (Internet of Things) and IoS (Internet of Services) link industry via its network of the supply chain and beyond the industrial sector while the cyber-physical networks allow "intelligent" industry.

Moreover, other writers point out the close relationship between industry 4.0, the CPS and IoT. Liao and others. The Systematic Industry Analysis 4.0 found 249 papers published by July 2016. Just three documents for IoT and CPS for Industry 4.0 were not included.

In this regard, our scenario studies consider CPS and IoT to be essential technological components of the basic components of the concept of industry 4.0. The fourth industrial revolution, however, includes some manufacturers, more or less developed concurrently with the CPS and IoT technologies, which play a vital role in increasing the features of the new production mechanism. Innovative products for industry 4.0 were also included in big data, RFID, cloud and additives (three diagrams) and blockchain. But it was not complete.

IoT is the worldwide relationship of physical objects with sensors, actuators, and communication technologies, according to Internet. Cloud computing is a combination of current technology and the requisite gene outsourcing benefits, the key feature of which is virtualization of compute resources and services. In order to encourage new applications and improve present applications, IoT may use a wide range of areas, such as manufacturing, health, transport and electricity.

7.3.3 SUSTAINABLE ENVIRONMENT AND INDUSTRY 4.0

In order to assess sustainable development opportunities from the microwave or macro perspective, the authors conducted a cutting-edge analysis through horizontal integration across the entire value creation network. Industry 4.0 offers significant chances of achieving sustainable manufacturing (Kiel et al., 2017). The result was that Industry 4.0 presents an excellent opportunity to achieve an intelligent cross-link of sustainable industrial value from the point of view of a life cycle.

Beier et al. said that there has been very limited research focusing on the influence of the digital sector on relevant aspects of industry sustainability. The authors analysed the anticipated environmental and social sustainable impacts of Industry 4.0 on German and Chinese companies. The environmental factor has been measured using materials, energy efficiency, ability to incorporate renewable energy sources and the implementation of environmental policies or standards. The findings show that the change can have a positive impact on environmental issues and increase social challenges.

The extensive digitalisation provided in Industry 4.0, by enhancing precision, performance, real-time external environmental accounting and Burritt and Christ environmental management accounting, can have a positive effect on corporate sustainability. The authors understand how to integrate the more widespread and previously undeveloped corporate sustainability context into this agenda.

Through a research methodology that focuses on semi-structured interviews between experts from 46 manufacture firms from the three leading German industry (DF) Kiel et al. argue that Thing's industrial internet (i.e., IoT) opportunities and challenges are not well known. The authors discuss the implications for TBL. The structural and inductive review reveals that the benefits and challenges of IIoT span all three areas of sustainability. Productivity of capital is the benefit for companies closely related to the environment.

The Dutch and others. Study aims, through comparative policy analyses through China and Taiwan, to expose competition and coalition trends and anatomize cross-sector policy materials of Industrie 4.0, with a view to providing guidance for sustainability in the sense of Industry 4.0. The aim is to provide a precise national analysis tool that promotes government technology and innovation. Both countries are concentrating on environmental policy, so that the necessary infrastructure can be developed by Industry 4.0.

Finally, we explored a number of previous literary strands linking Industry 4.0 with a sustainable setting. In short, previous studies have used many approaches, including Ford and Despeisse exploratory studies. There have been interviews with experts in the area. We used the analyses of material and statistical evidence. In conclusion, after all of these studies have been reviewed, barriers to and opportunities for the introduction of Industry 4.0 remain unsure and the related advances have not been adequately studied in terms of environmental safety because they are still new technologies (Petter and Per 2009). In order to minimise pollution in the environment and achieve sustainability, the 4R – reduce, reuse, recycle and rebuild – may be used (Verma and Sharma 2017). Therefore, sustainability and eco-innovation will be introduced in Industry 4.0 and environment. Besides the writers' contributions, this is a new discovery.

7.4   SIX SIGMA

In addition to a strong lean approach mainly focused upon disposal of waste (muda), an equally successful way is most frequently used to delete or change the process to eliminate mura incoherence (Miguélez et al., 2014). The global implementation of Six Sigma demonstrates its important role in encouraging the success of companies that many companies confirm (Valles et al., 2009). In the first years of Motorola in the 1980s the concept was introduced and further developed by General Electric (GE) and AlliedSignal later in the 1990s (Braunscheidel et al., 2011). Since then, the Six Sigma concept has been renowned and has been implemented globally by many businesses, including Fortune 500 (Goh, 2002) as it contributes to significant changes in both operational as well as financial performance and customer satisfaction through the minimisation of comprehensible products and services (Garza-Reyes, 2015). For this, it was well-known.

Previously, no final conclusions relevant to the financial performance of quality assurance practises were found (Powell et al., 2017). Many companies of Six Sigma are engaged in financial and accounting practises, in particular in identifying Six Sigma (Pyzdek & Keller, 2014), financial benefits that distinguish Six Sigma from other quality management methods (Pande & Holpp, 2002).

7.5   SIX SIGMA AND INDUSTRY 4.0

Following an investigation on sustainable manufacturing factors such as limitations on the supply chain and market pressure, Fargani et al. (2016) suggested using Lean and Six Sigmas as a way of achieving sustainable production. Nagalingam et al. (2013) introduced a Six Sigma Methodology system for estimation of the efficiency of product recovery and design returns of ID OVI. Zhang and Awasthi (2014) have implemented another system with an emphasis on sustainability leadership – a combined six-sigma process. On the side of energy conservation, Chugani et al. (2017) confirm that six sigmas can serve as a way to handle energy usage efficiently and advise organisations to include it in their policies.

Jayaram (2016) pointed to the joint addition of Industry 4.0 and Lean Six Sigma and introduced an overall supply chain management model. An algorithm to predict the robust manufacturing process to synthesise the tolerance that is an introduction to the process improvement in the Giannetti & Ransing (2016) and Giannetti (2017) industry of six sigma applications. In order to rectify the deviation in real time manufacturing processes using Industry 4.0, Eleftheriadis and Myklebust (2016) apply a procedure. Basios and Loucopoulos (2017) proposed an approach for the management of the data collected by the use of Industry 4.0 technologies to help businesses in taking strategic decisions. The Dogan and Gurcan (2018) outlined data processing methodologies for the six sigma-related Industry 4.0 data collection phases. In maintenance, Antosz and Stadnicka (2018) suggested on-site maintenance data input via six sigma-based decision-making procedures for the maintenance phase of the Cyber Physical System (CPS). DMAIC's sub-methodology for IoT (interne of subjects) ventures was derived by writers in Hammad et al. (2017) for a predictive case study.

7.6   CASE STUDY

DESCRIPTION

In this article, AMPLiFII is the case study. This is a four-year project for Innovate UK, with AMPLiFII 1 followed by AMPLiFII 2 closing in early 2020. It was intended to build a battery pilot line, especially battery modules that use cylindrical cells. This project was planned. The batteries can be configured in size and capacity by changing the cells in use or the series and parallel connexion in this module. The use of mobile controls and the use of the RPAW pulse arc welder known by its cell loading system or CLS is important for on-line automation.

The case study uses the CLS example of cell voltage. The CLS regulates the cell's inner strength and stress. The top and bottom acceptation limits shall be set manually by the cell supplier listed or the assembled module design specification. A research was performed on nest cells with 30 cells each. In the past, the cells that have been rejected have been quarantined and can be destroyed. The new monitoring method for collecting samples of CLS data offers an insight into the quality of this component and data have been processed to avoid confusion between process performance measurement and batch performance measurement.

7.7   METHODOLOGY

In order to discuss Industry 4.0 and environmental sostenibility, we have studied and analysed theoretical and literary analyses by various scholars. Our study carried out a literature review and used the method of gathering all manuscripts on the topic of Environmental Sustainability 4.0 written from 2000 to 2020. Secondary data was also used. It has enabled us to analyse from the viewpoint of environmental protection the positive and negative impacts of Industry 4.0 on the production sector and key technologies. The production method, i.e., raw materials, resources and expertise, needed to process inputs in goods, waste and end-of-life products, and greenhouse gas (GHG) emissions was passed for the first time in order to achieve a comprehensive design (McWilliams et al., 2016).

Value stream mapping in any device is one of the best approaches to find waste (Keskin et al., 2013 & Badurdeen and Faulkner, 2014). In the current study, the researchers modified the **VSM** into **EVSM** by integrating and analysing the power components in terms of time (Müller et al., 2014). In order to assess opportunities for energy saving (Erlach and Westkämper 2009), the EVSM considers energy use and waste level in operation (Herrmann et al., 2013, Verma & Sharma 2016 and Verma & Sharma 2019). Not only for the diagnosis, but also for energy budgeting and conservation steps should the suggested template be used (Chatterjee 2014). The EVSM is used with six-sigma methods in a robust energy efficiency process (Michael et al., 2014).

7.7.1 IMPLEMENTATION OF LEAN-SIX-SIGMA FOR PROCESS OPTIMIZATION

The six-sigma project starts with the identification of CTQ (critical to quality). In the present case, the CTQ is to minimize the energy conservation in the existing system. The tools used for the DMAIC in the selected process flow (Figure 7.5) are as follows:

- *Define* - The method, the customer's voice and their needs, **(SIPOC Diagram)**
- *Measure* - The current process and the related data are collected; measure process capability 'as-is'. **(Current value stream mapping)**
- *Analyze* - Data to assess and measure causal-effect relationships. Identify relationships and aim to take into account all aspects. **(Fishbone diagram, Time study).**
- *Improve* - The current data analysis method uses techniques such as experimental design, error evaluation and standard work to develop a new, future state method. **(Future Value stream map)**
- *Control* - The future state mechanism ensures that any deviations from the target are rectified before errors occur.

7.7.1.1 Define

The first step is to define the Critical to Quality (CTQ) and the Critical to Quality is to identify and eliminate the energy-consuming steps. To identify the Critical to Quality six-sigma tool SIPOC (suppliers, inputs, process, outputs, and customer) is used. This is a visual aid in every company to record processes from start to finish. Just before work starts, the SIPOC diagram describes all the elements in the

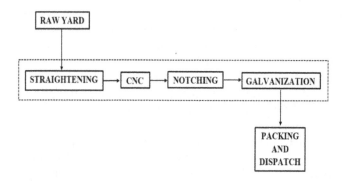

FIGURE 7.5 Layout of selected process.

development phase for the project. Figure 7.6 of the SIPOC helps to identify a complex project that can not be measured correctly and is usually used in the Six Sigma DMAIC technique calculation process.

7.7.1.2 Measure

EVSM is an efficient method for calculating the amount of waste of energy in-operation. The process starts with identifying the chronological order of the processes involved and other relevant information's Figure 7.7

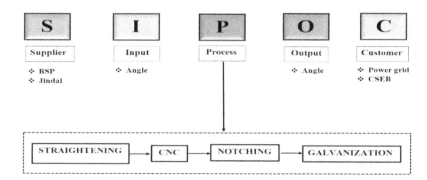

FIGURE 7.6 Map SIPOC.

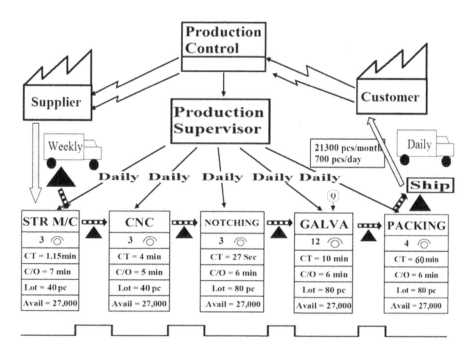

FIGURE 7.7 Information flow value stream mapping.

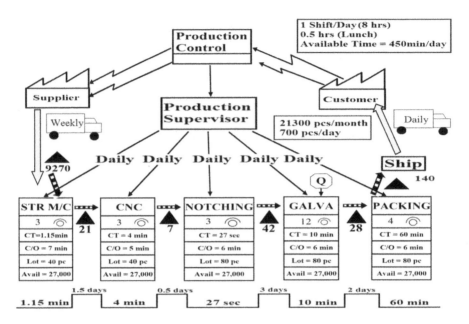

FIGURE 7.8 Mapping of current stream value.

The steps involved in manufacturing are, straightening (STR M/C), CNC (cutting, punching, drilling, stamping), Notching and Galvanizing (GALVA).

The information's, which are added in VSM is, Machine name, Energy (E), Cycle time (CT), Change over time (C/O), Number of workers engaged in the machine, Pieces per lot, CVSM shown in Figure 7.8 from these data current VSM is drawn Information's which are extracted for the data are:

- Overall production (angle & plate both) per annum = 6408 metric ton
- Overall production of angle (included cleats) per annum = 5890 metric ton
- Therefore, the Number of angles produce is 21300 pieces/month.

7.7.1.2.1 Create Energy Value Stream Mapping

The energy consumption is based on energy consumed every day, month and year during each process for each added value and non-value added operation. The energy value mapping is shown in Figure 7.9.

Timeline Diagram

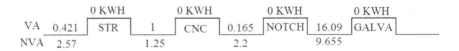

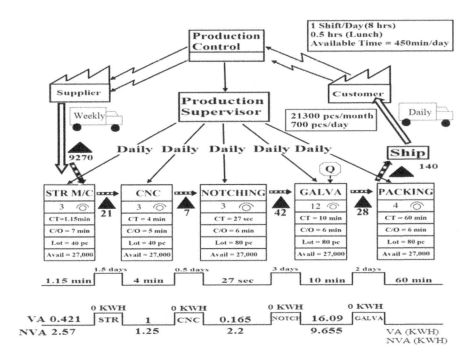

FIGURE 7.9   Energy value stream mapping.

Calculation of energy consumption by value added (EVA)& non-value-added activity (ENVA):

EVA = 17.677 kWh

ENVA = 15.672 kWh

7.7.1.2.2 Time Study

The cycle time is measured by using a stopwatch for every strip. Depending on the area or operation of bars, the cycle time varies. The process starts when the bar is loaded on the moving table and stops after unloading it. Approx. 1.15 min takes for 1 angle of, hence the process will continue for obtaining the average cycle.

With a stopwatch, the time of every operation is measured. With the aid of a stopwatch, all operations are carried out. The time of the loop is from beginning to end. The process starts with the loading of the bar on the machine and time are noted for the process like stamping, punching, and cutting and unloading. The data are collected many times and the average time for further analysis is reported.

The CNC cycle time, emergency operations and galvanisation lists Table 7.4, Table 7.5 and Table 7.6.

7.7.1.3   Analysis of Time Study

Over-processing is among seven wastes of lean manufacturing which summarizes the excess time and effort spend into product processing. Practitioners emphasize to optimize the process with respect to minimizing the wastage of time and materials.

TABLE 7.1
Energy Value Stream Mapping Data

Process/Data	Straightening	CNC	Notching	Galvanization
Energy (KW)	22	15	22	96.55
C/T	1.15 min.	4 min.	27 sec.	10 min.
C/O	7 min.	5 min.	6 min.	6 min.

It has been identified that the main problem occurred in Galvanizing Section is. "Over-processing". Applying the protective zinc coating against rusting is a galvanising process. The fishbone diagram has been developed to define over-processing cause(s) (Figure 7.10).

    i. Man:
        There is a plethora of manpower in this industry, which is necessary for continuous production for various processes. The working hours are compiled in the shifts. Lack of coordination has been found in many cases during shift change and lunch breaks hours. Also, less-skilled worker and inefficient subordinate, are being the other reasons of over-processing.

    ii. Machine:
        Proper maintenance management of machine is required for quality product with minimum rejection and rework. Proper handling and adequate maintenance management of machine processing such as for zinc coating section (for the considered industry) etc. will lead to reduction of rejection and rework. Apart from these, scheduled maintenance, machine set-up & calibration is also the area of concern for zero breaks down and minimal wastage.

    iii. Material:
        Good management of raw materials helps the operators for continuous and smooth working. Any bottleneck or blockage (due to any reason) leads to piling of inventory and reduction in productivity. It is observed that, the manufacturing unit lacks the proper material handling system for heavy and bulky components and the same is being carried out manually. This is not only consuming time but also hazardous and affects the production rate. Also, the inventory storage is not proper, the raw materials are kept unprotected in open space which causes rusting. In order to remove the rusting, materials are kept for some time in the pickling tank which increases the processing time in the production flow (Figure 7.11).

    iv. Method
        For the pickling process, First Come first Out method is not followed which causes some racks to remain dipped in the pickling tanks due to negligence of the operator. This creates a loss in time and capital because that product is

TABLE 7.2
Energy Calculation for Different Processes

Process / Energy	Straightening	CNC	Notching	Galvanization
EVA (C/T)	$22 \times \frac{1.15}{60} = \mathbf{0.421 KWH}$	$15 \times \frac{4}{60} = \mathbf{1 KWH}$	$22 \times \frac{27}{3600} = \mathbf{0.165 KWH}$	$96.55 \times \frac{10}{60} = \mathbf{16.09 KWH}$
ENVA (C/O)	$22 \times \frac{7}{60} = \mathbf{2.57 KWH}$	$15 \times \frac{5}{60} = \mathbf{1.25 KWH}$	$22 \times \frac{6}{60} = \mathbf{2.2 KWH}$	$96.55 \times \frac{6}{60} = \mathbf{9.655 KWH}$

TABLE 7.3
Time Study for Straightening (STR M/C)

S. No	Cycle	Start Time	End Time	Total Time	Remark
1.	CYCLE 1	02:21:00	02:22:00	00:01:15	
2.	CYCLE 2	02:23:00	02:24:00	00:01:25	
3.	CYCLE 3	02:24:00	02:25:00	00:01:20	
4.	CYCLE 4	02:26:00	02:27:00	00:01:21	setup time for adjusting steering
5.	CYCLE 5	02:28:00	02:29:00	00:01:04	worker was busy for inspection. (takes 4 min)

TABLE 7.4
Time Study for CNC (CNCVP-942)

S.No	Cycle	Start Time	End Time	Total Time	Remark
1.	CYCLE 1	02:36:00	02:40:00	00:04:00	6 min loss for checking machine setup
2.	CYCLE 2	02:41:00	02:45:00	00:04:00	35 sec loss for changing the punching pin
3.	CYCLE 3	02:45:00	02:49:00	00:04:00	
4.	CYCLE 4	02:56:00	02:59:00	00:03:00	
5.	CYCLE 5	03:02:00	03:05:00	00:03:00	

TABLE 7.5
Time Study for Notching (Notch 3)

S. No	Cycle	Start Time	End Time	Total Time	Remark
1.	CYCLE 1	02:21:00	02:22:00	00:00:26	
2.	CYCLE 2	02:23:00	02:24:00	00:00:25	
3.	CYCLE 3	02:24:00	02:25:00	00:00:27	
4.	CYCLE 4	02:26:00	02:27:00	00:00:27	2 time change over Taking 13 min
5.	CYCLE 5	02:28:00	02:29:00	00:00:27	Worker was busy in chip collection. (Takes 9 Min)

over-processed in the respective tank. In the process section, there is no standard time of dip or standard process that has been found in this industry and the process completely depends upon the operator's own observations and work experiences.

TABLE 7.6
Time Study for Galvanization Process

Cycle	Start Time	End Time	Total Time (Min)	Dipping Time (Min)	Temp (°C)	Weight (Kg)	Remark
CYCLE 1	9:21:00	9:31:00	10.01	3.25	453	3174	
CYCLE 2	9:37:00	9:46:00	8.34	3.30	452	4168	Zinc Addition Takes (7 Min)
CYCLE 3	9:47:00	10:02:00	14.48	1.53	448	2474	Nickle Powder Addition Used for Reducing Gases Takes (6 Min)
CYCLE 4	10:14:00	10:21:00	6.37	6.37	455	4304	
CYCLE 5	10:25:00	10:35:00	10.01	4.42	450	2914	
CYCLE 6	10:36:00	10:45:00	9.11	3.28	451	4298	
CYCLE 7	11:24:00	11:35:00	10.13	3.53	455	2912	
CYCLE 8	11:37:00	11:47:00	10.47	3.45	452	4849	
CYCLE	**Start Time**	**End Time**	**Total Time (Min)**	**Dipping Time (Min)**	**Temp (°C)**	**Weight (Kg)**	**Remark**
CYCLE 9	11:49:00	12:01:00	11.01	4.34	447	3493	
CYCLE10	12:04:00	12:14:00	10.51	4.21	447	4169	After dipping the rack goes through back to the dryer for 3.50 min because of an insufficient area on quenching one.

7.7.1.4   Improvement

7.7.1.4.1 Future State Map

A potential map links the distance between present and ideal. It stresses the limitations of technological constraints, budgets and time. The aim of value stream mapping is to identify and remove waste sources and enforce the future government value stream so that it can be implemented. Symbol used in future VSM Shown in Figure 7.12. The primary objective of the future state map is to create a supply chain where the individual processes are connected either by constant streaming or pulling to their customers and where each process gets closer to when a customer wants the product.

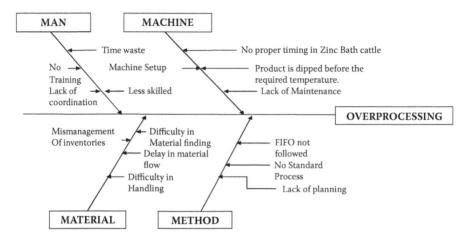

FIGURE 7.10 Fishbone diagram.

FIGURE 7.11 Mishandling of products.

7.7.1.4.2 Symbols Used in Future State Map
7.7.1.4.3 Analysis of Future State Map

- During the conversion from the inbound to the outbound logistics, the processing steps are the various operations performing on the commodity A production Kanban is applied between these two stages to ensure an uninterrupted and in time supply of raw materials. FVSM shown in Figure 7.13.
- The manufacturing method Kanban and material pull is used between stock and process line. The material pull system helps to reduce in-process inventory and maintain the smooth flow of material.
- Between Straightening and CNC, a supermarket is introduced. Between supermarket and straightening, a production Kanban is used & between CNC and supermarket, a withdrawal Kanban is used. This production Kanban will give the

SYMBOL	NAME OF THE SYMBOL	DESCRIPTION
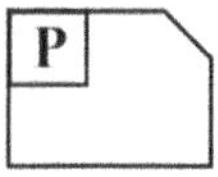 P	PRODUCTION KANBAN	Manufacture Kanban shall define the product form and volume to be produced by the previous procedure.
	Kanban Withdraw	Withdrawal Kanban shows the form and quantity of goods that are supposed to exit from a previous phase of manufacturing.
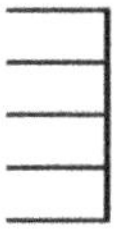	SUPERMARKET	A supermarket is a form of inventory management whereby different components are stored without knowing the order in which the components are removed from the inventory.
	KAIZEN	Kaizen symbol is used to highlight improvements needed.
	MATERIAL PULL	This pull sign is the physical removal from the supermarkets of the stored inventory.
	IMPROVEMENT/ SOLUTION	The cloud symbol is used to illustrate the concepts or solutions suggested.
—FIFO→	FIFO	This icon is a device that restricts inventory input to the first-in-first-out.

FIGURE 7.12 Future state map symbols.

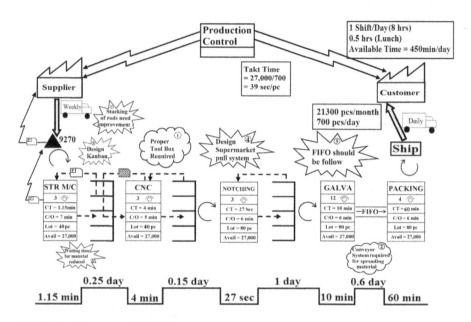

FIGURE 7.13 Future flow of state importance.

information about the product requirement of the CNC while the withdrawal Kanban will give the information of the products that have been withdrawn from the supermarket to ensure the flow of material without interruption.

- Between CNC and notching, supermarket and material pull system is applied. The supermarket is the stock control where different components can be retained without understanding in what order the components are removed from the stock and the pull sign is a physical removal of the product from supermarkets.
- Notching is followed by the galvanization process. Between notching and galvanization, supermarket and material pull system is being used. The quality problem symbol indicates a quality problem in the galvanization process. In order to limit the inventory input, the FIFO symbol which indicates first in first out system is applied between galvanization and packing. Pullover is performed between packing and shipping to ensure the outlet of the product as soon as its production is completed.
- The process information box will be added with details. The knowledge flow occurs in the VSM. The information that has been added in future state VSM. In the future state map, kaizen symbols have been used to highlight the improvements needed. Following are the information carried by kaizen symbols in future state map:
  i.   The first kaizen is inserted between supplier and inventory that tells that the stacking of raw materials needs improvisation.
  ii. The second kaizen is used between straightening and CNC indicates the improvement needed to reduce the waiting time.

iii. Third kaizen is used between straightening and CNC which indicates the need for the introduction of the Kanban system between these two.

iv. Fourth kaizen is introduced between CNC and notching, indicates the requirement of supermarket between CNC and notching.

v. Fifth kaizen is used between galvanization and packing, which indicates the need for the introduction of the FIFO system.

- Using the cloud symbol or solution symbol, suggestions or ideas will be highlighted. The following ideas or solution are suggested based on observations and future state map:

  i. A proper toolbox along with its proper arrangement is required in CNC.

  ii. Conveyor system can be operated for the proper spreading of the material.

- The final schedule is generated which shows time to wait and process. This can be used for Lead Time and Cycle Time measurements.

7.7.1.5   Control

The final schedule is generated which shows time to wait and process. This can be used for lead time and cycle time measurements. Measures taken for the improvements and control of different problems between the processes are:

1. **Control of excess inventories between the two workstations**:

"Supermarket symbol" is introduced between two workstations. This supermarket indicates the management of inventory between the two workstations in which various parts can be kept without knowing in what order the parts will be taken from the inventory.

- "Production Kanban" & "withdrawal Kanban" is used. This production Kanban will give the information about the product requirement of the CNC & ensure the flow of material at the time of requirement without delay. Withdrawal Kanban will leave the information of the products that have been withdrawn from the supermarket. This is to ensure the flow of material without interruption.

2. **Control of over-processing in galvanization**:
   - To control quality problem, need to ensure that, high tensile and mild steel should be dipped separately in different lots, because if these two are dipped together then one of them will undergo over-processing.
   - Need to control the processing of that type of product, having a material defect, because this creates an obstacle between the continuous material flow. Inspection of material defects should be done in the upstream process, not after galvanization. This inspection program reduces the rework problem.
   - There are some problems(like black spots) are identified after pickling process & after dryer which can be eliminated by grinding of material

before dipping in zinc bath cattle, this action reduces the rework problem and save that amount of zinc that is wasted for rework and hence reduces the zinc consumption cost which is beneficial for the industry.

3. First-in first-out system is applied between galvanization and packaging:

Now between galvanization and packing FIFO symbol that indicates first in first out system is applied in order to limit the inventory input.

4. Control of delay time in shipment after packaging:

Pullover is applied between packing and shipping to ensure the shipment of the product as soon as its production is completed.

5. Introduction of KAIZEN:

KAIZEN symbol indicates improvement done or needed.

6. Introduction of cloud symbol:

The cloud symbol shows suggestions that improve the existing process.

7.8   RESULT AND DISCUSSION

After the drawing and review of the current value stream map, a future state map was drawn and all changes were made, it was found that the delay between different operations was reduced (Table.7.7)

In the energy estimate, 17677 kWh energy has been found to be value-added and 15672 kWh energy is applied as non-value. The relation between the current state and future state stream values mapping is shown in Figure 7.14.

TABLE 7.7
The Reduction in Delay Time Between Various Operations

Delay between operations	Delay time of current VSM (Days)	Delay time of future VSM(Days)
Straightening – CNC	1.5	0.25
CNC – Notching	0.5	0.15
Notching – Galvanization	3	·1
Galvanization – Packing	2	0.6

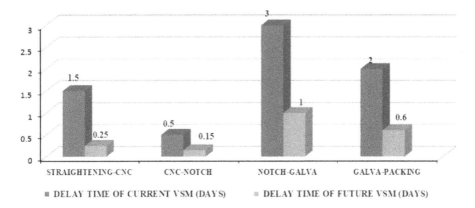

FIGURE 7.14 Comparison of delay time between current state & future state value stream map.

7.8.1 DETERMINATION OF BOTTLENECK STATION

Figure 7.15 shows the bottleneck in the process line. The graph and cycle time of different operations have been drawn. The time of takt is a dotted line. The time for tact is the time period for manufacturing a product to fulfil the customer's requirements. The Figure 7.15, It has been observed that the maximum deviation from the takt time is shown in the galvanization operation. Hence, the bottleneck occurs in the galvanization process.

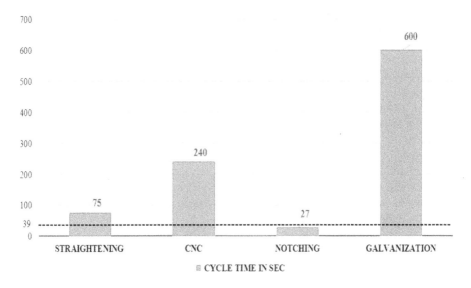

FIGURE 7.15 Cycle time & takt time.

$$Takt\ Time = \frac{\text{Available working time per shift}}{\text{Customer demand rate per shft}}$$

7.9   OUTCOME CRITERIA

In the case study Large scale industry "**KALPATARU POWER TRANSMISS-ION LIMITED",** The methodology for the measurement of denial, revision and rework with non-value added energy was introduced in the overall process line in accordance with value stream mapping. The processes for straightening, CNC, coding and galvanization have been taken into account in this value stream map. The time lag between stations is decreased after the current and future state value stream map has been drawn. It also highlighted the over-processing issue while the stacking issue was addressed in the inventory. The method used is the FIFO method, i.e., firstly to retain the material flow in industry. In the future state diagram, detailed changes are indicated as Kaizen.

7.10   CONCLUSION

We have shown the inputs in stage one (deployment scenario) in industry 4.0 which permits automation, digitalisation and integration in the next steps and scenarios. In the second stage (operating scenario), the processing of raw materials into goods takes place using real-time data and custody and thus intelligent development. Step three includes the partnership between Industry 4.0 and the SDG, which will ensure long-term sustainableness in the fourth phase of the study. Although it will not be included in a quantitative impact evaluation, it should be noted that the environmental implications of the research are not part of that. However, well-established policies to protect the environment and its habitats need to be enforced from the negative impacts of Industry 4.0 on the energy transfer of policy mix instruments, to achieve full benefits of industry 4.0. For all stakeholders, this should be an problem. The policies in harmony with the technology and Industry 4.0 as well as energy saving can be reflected in the sense of production, transportation and consumption for a sustainable future. This method, which is linked to the third word, has provided proof of positive qualitative impacts. For the quantitative environmental impact assessment, the effect of Industry 4.0 technology must be analysed through various methodologies such as material strength, life cycle analysis and energy estimates. In order to follow such a structure, a context and time frame need to be defined, although this is outside our research reach.

REFERENCES

Antosz, K., and Stadnicka, D. (2018). Possibilities of maintenance service process analyses and improvement through six sigma, lean and industry 4.0 implementation. In *IFIP Advances in Information and Communication Technology* (Vol. 540). Springer International Publishing. https://doi.org/10.1007/978-3-030-01614-2_43

Badurdeen F., and Faulkner W. (2014) Sustainable Value Stream Mapping (Sus-VSM): Methodology to Visualize and Assess Manufacturing Sustainability Performance. *Journal of Cleaner Production*, 85, pp. 8–18.

Basios, A., and Loucopoulos, P. (2017). Six Sigma DMAIC Enhanced with Capability Modelling. 2017 IEEE 19th Conference on Business Informatics (CBI), 02, 55–62. https://doi.org/10.1109/CBI.2017.70

Baswaraj A., Rao M., and Kumar A. (2015) Energy consumption Optimization for Secondary Steel by Six Sigma, *International Journal of Modern Trends in Engineering and Research*, 2, 1690–1695.

Behera, J. (2015). Energy consumption and economic growth in India: A reconciliation of Disaggregate Analysis. *Journal of Energy Technologies and Policy*, 5(6), pp. 15–27. Retrieved from www.iiste.org

Bertoldi P., and Mosconi R. (2020) Do Energy Efficiency Policies save Energy? A New Approach Based on Energy Policy Indicators (in the EU Member States). *Energy Policy*, 139, pp. 1–18.

Bettoni L., Mazzoldi L., and Ferretti I. (2015) Integrated Energy Value Analysis a New Approach. International Conference on Advances in Production Management Systems: Innovative Production Management towards Sustainable Growth, pp. 670–679.

Bonilla S., Silva H., Terra M., Franco G.R., and Sacomano J. (2018) Industry 4.0 and Sustainability Implications: A Scenario-Based Analysis of the Impacts and Challenges. *Sustainability*, 10, pp. 3740.

Braunscheidel, M. J., Hamister, J. W., Suresh, N. C., & Star, H. (2011). An institutional theory perspective on Six Sigma adoption. *International Journal of Operations & Production Management*, 31(4), pp. 423–451. https://doi.org/10.1108/01443571111119542

Bunse K., and Vodicka M. (2011) Integrating Energy Efficiency Performance in Production Management Gap Analysis between Industrial Needs and Scientific Literature. *Journal of Cleaner Production*, 19, pp. 667–679.

Chatterjee R. (2014) Value Stream Mapping Based on Energy and Cost System for Biodiesel Production. *International Journal of Sustainable Manufacturing*. 3, pp. 95–115.

Chugani, N., Kumar, V., Garza-Reyes, J. A., Rocha-Lona, L., and Upadhyay, A. (2017). Investigating the green impact of Lean, Six Sigma and Lean Six Sigma. *International Journal of Lean Six Sigma*, 8(1), pp. 7–32. https://doi.org/10.1108/IJLSS-11-2015-0043

Da Silva F.S.T., da Costa C.A., Crovato C.D.P., and da Rosa Righi R. (2020) Looking at Energy through the Lens of Industry 4.0: A Systematic Literature Review of Concerns and Challenges. *Computers & Industrial Engineering*, 143, pp. 1–21.

Dogan, O., and Gurcan, O. F. (2018). Data perspective of lean six sigma in industry 4 .0 era: A guide to improve quality. Proceedings of the International Conference on Industrial Engineering and Operations Management, 2018(JUL), 943–953.

Eleftheriadis, R. J., and Myklebust, O. (2016). A guideline of quality steps towards zero defect manufacturing in industry. Proceedings of the International Conference on Industrial Engineering and Operations Management, 332–340.

Erlach K., and Westkämper E. (2009) Energy Value Stream – The Path to the Energy Efficient Factory. *Procedia CIRP*, 17, pp. 368–373.

Fargani H., Cheung W. M., and Hasan R. (2016). An empirical analysis of the factors that support the drivers of sustainable manufacturing. *Procedia CIRP*, 56, 491–495. https://doi.org/10.1016/j.procir.20 16.10.096

Fu B., Shu Z., and Liu X. (2018) Blockchain Enhanced Emission Trading Framework in Fashion Apparel Manufacturing Industry. *Sustainability*, 10, Art. no. 1105.

Garza-Reyes, J. A. (2015). Green lean and the need for Six Sigma. *International Journal of Lean Six Sigma*, 6(3), pp. 226–248. https://doi.org/10.1108/IJLSS-04-2014-0010

Giannetti, C., and Ransing, R. S. (2016). Risk based uncertainty quantification to improve robustness of manufacturing operations. *Computers and Industrial Engineering*, 101, 70–80. https://doi.org/10.1016/j.cie.2016.08.002

Giannetti, C. (2017). A framework for improving process robustness with quantification of uncertainties in Industry 4.0. 2017 IEEE International Conference on INnovations in Intelligent SysTems and Applications (INISTA), 189–194. https://doi.org/10.1109/INISTA.2017.8001155

Goh, T. N. (2002). A strategic assessment of Six Sigma. *Quality and Reliability Engineering International*, 18(5), pp. 403–410. https://doi.org/10.1002/qre.491

Hammad, M., Khan, A. K., Deng, S., Rashid, A., Khan, J. A., and Zulfiqar, F. (2017). An integration of Kano Model, QFD and Six Sigma to present a new description of DFSS . *European Journal of Business and Management Online*, 9(6), 2222–2839.

Herrmann C., Bogdanski G., Schönemann M., Thiede S., and Andrew S. (2013) An Extended Energy Value Stream Approach Applied on the Electronics Industry. Proceedings of APMS 2012, pp. 65–72.

Hermann, M., Pentek, T., and Otto, B. (2016). Design Principles for Industrie 4.0 Scenarios. 2016 49th Hawaii International Conference on System Sciences (HICSS), 3928–3937. https://doi.org/10.1109/HICSS. 2016.488

Jayaram, A. (2016). Lean six sigma approach for global supply chain management using industry 4.0 and IIoT. 2016 2nd International Conference on Contemporary Computing and Informatics (IC3I), 89–94. https://doi.org/10.1109/IC3I.2016.7917940

Kagermann, H., Lukas, W., and Wahlster, W. (2016). Abschotten ist keine Alternative. VDI Nachrichten, (16).

Kane J., and Dupont E.I. (2003). Using Six Sigma to drive energy efficiency improvements at DuPont. American Council for an Energy-Efficient Economy, 50–61. https://aceee.org/files/proceedings/2003/data/papers/SS03_Panel2_Paper07.pdf

Kaushik P., and Khanduja D. (2007) DM Make Up Water Reduction in Thermal Power Plants Using Six Sigma DMAIC Methodology, *Journal of Science and Industrial Research*, 67, 36–42.

Kaushik P., Grewal C., Bilga P., and Khanduja D. (2008) Utilising Six Sigma for energy conservation: A process industry case study, *International Journal of six sigma of competitive Advantage*, 4(1), pp. 18–34.

Keskin C., Umut A., and Gulgun K. (2013) Value Stream Maps for Industrial Energy Efficiency. *Assessment and simulation tools for sustainable energy systems*. London: Springer, pp. 357–379.

Kiel D., Müller J.M., Arnold C., and Voigt K.-I. (2017) Sustainable Industrial Value Creation: Benefits and Challenges of Industry 4.0. *International Journal of Innovation Management*, 21, Art. no. 1740015.

Krautkraemer, J. (2005). Economics of Natural Resource Scarcity: The State of the Debate. Discussion Papers, April.

Kumar M., and Antony J. (2008) Comparing the quality management practices in UK SMEs. *Industrial Management & Data Systems*, 108, pp. 1153–1166.

Lee J., Yuvamitra K., and Kozman T. (2014) Six Sigma Approach to Energy Management Planning. *Strategic Planning for the Energy and Environment*, 33(3), pp. 23–40.

Lee J., Yuvamitra K., Guiberteau K., and Kozman T. (2014) Six Sigma Approach to Energy Management Planning. *Strategic Planning for the Energy and Environment*, 33, pp. 23–40. Association of Energy Engineers.

Liao Y., Deschamps F., Loures, E.D.F.R., and Ramos L.F.P. (2017) Past, Present and Future of Industry 4.0 - A Systematic Literature Review and Research Agenda Proposal. *International Journal of Production Research*, 55, pp. 3609–3629.

Lieder, M., and Rashid, A. (2016). Towards circular economy implementation: a comprehensive review in context of manufacturing industry. *Journal of Cleaner Production*, 115, 36–51. https://doi.org/https:/ /doi.org/10.1016/j.jclepro.2015.12.042

Lin K., Shyu J., and Ding K. (2017) A Cross-Strait Comparison of Innovation Policy under Industry 4.0 and Sustainability Development Transition. *Sustainability*, 9, Art. no 786.

McWilliams A., Parhankangas A., Coupet J., Welch E., and Barnum D.T. (2016) Strategic Decision Making for the Triple Bottom Line. *Business Strategy and the Environment* , 25, pp. 193–204.

Michael D., Martin B., Simon S., and Franke J. (2014) A New Approach to Integrate Value Stream Analysis into a Continuous Energy Efficiency Improvement Process. 39th Annual Conference of the IEEE Industrial Electronics Society, Vienna, Austria, pp. 7502–7507.

Miguélez, M.S., Errasti, A., and Alberto, J.A. (2014). Lean-Six Sigma Approach put into Practice in an Empirical Study. In: Prado-Prado J., and García-Arca J. (Eds) *Annals of Industrial Engineering 2012*. London: Springer. https://doi.org/10.1007/978-1-4471-5349-8_30

Müller J.M., Kiel D., and Voigt K.-I. (2018) What Drives the Implementation of Industry 4.0? The Role of Opportunities and Challenges in the Context of Sustainability. *Sustainability*, 10, pp. 247.

Müller E., Stock T., and Schillig R. (2014) A Method to Generate Energy Value-Streams in Production and Logistics in Respect of Time-and Energy-Consumption. *Production Engineering Research and Development*, 8, pp. 243–251.

Murugananthan V., Govindaraj K., and Sakthimurugan D. (2014) Process Planning through Value Stream Mapping Foundry. *International Journal of Innovative Research in Science*, 1, pp. 1140–1143.

Nagalingam, S. V., Kuik, S. S., and Amer, Y. (2013). Performance measurement of product returns with recovery for sustainable manufacturing. *Robotics and Computer-Integrated Manufacturing*, 29(6), pp. 473–483. https://doi.org/10.1016/j.rcim.2013.05.005

Petter S., and Per G. (2009) Concepts for Simulation Based Value Stream Mapping. Proceedings of the 2009 Winter Simulation Conference (WSC). IEEE. pp. 2231–2237.

Powell, D., Lundeby, S., Chabada, L., and Dreyer, H. (2017). Lean Six Sigma and environmental sustainability: the case of a Norwegian dairy producer. *International Journal of Lean Six Sigma*, 8(1), pp. 53–64. https://doi.org/10.1108/IJLSS-06-2015-0024

Preston, M., and Herron, J. P. (2016). Minerals and metals scarcity in manufacturing: The ticking time bomb. PwC. Available at: http://www.pwc.com/gx/en/services/sustainability/publications/metal-minerals scarcity.html.

Pyzdek, T., and Keller, P. (2014). *Six Sigma Handbook*, Fourth Edition. McGraw-Hill Education. https://www.accessengineeringlibrary.com/content/book/9780071840538

Rickand L., and Dugan J.P. (2008) Six Sigma from Products to Pollution to People. *Total Quality Management*, 19(1), pp. 1–9.

Siemens (2014a). Digital Factory Defects: A vanishing Species? Available at: http://www.siemens.com/innovation/en/home/pictures-of-thefuture/industry-and-automation/digital-factories-defects-a-vanishi ng-species.html

Sihag A., Kumar V., and Khod U. (2014). Application of Value Stream Mapping in Small Scale Industry. *International Journal of Mechanical Engineering and Robotics Research*, 3, pp. 738–746.

Soni Shashank, Mohan Ravindra, Bajpai Lokesh, and Katare S. K. (2013). Reduction of welding defects using six. *Reduction of Welding Defects Using Six Sigma Techniques*, 2(3), pp. 404–412.

Stock T., and Seliger G. (2016) Opportunities of Sustainable Manufacturing in Industry 4.0. *Procedia CIRP*, 40, pp. 536–541.

Swain M., Blomqvist L., McNamara J., and Ripple W.J. (2018) Reducing the Environmental Impact of Global Diets. *Science of the Total Environment*, 610, pp. 1207–1209.

Timothy G., Matthew S., Jeffrey B., and Jones A.J. (2011) Thermodynamic Analysis of Resources Used in Manufacturing Processes. *Article in Environmental Science Technology*, 43(5), pp. 1584–1590.

Valles A., Sanchezs, J. Noriega, S., and Nunez B.(2009) Implementation of Six Sigma in a Manufacturing Process, *International Journal of Industrial Engineering*, 16, pp. 171–181.

Verma N., and Sharma V. (Mar. 2015) "Lean Modelling – A Case Study For Indian SME" in *International Journal for Technological Research in Engineering.* 2(7), pp. 991–998, ISSN: 2347-4718.

Verma N., and Sharma V. (2016) Energy Value Stream Mapping a Tool to Develop Green Manufacturing. International Conference on Manufacturing Engineering and Materials, pp. 526–534.

Verma N., and Sharma V. (2017) "Sustainable Competitive Advantage by Implementing Lean Manufacturing—A Case Study for Indian SME". 4, pp. 9210–9217. Available Online at www.Sciencedirect.Com

Verma N., and Sharma V. (2019) "A Literature Review on Energy Value Stream Mapping (EVSM)" *International Journal of Advanced Science and Technology*, 27(1), pp. 1–8, ISSN: 2005-4238, SERSC Publication.

Wang, S., Wan, J., Zhang, D., Li, D., and Zhang, C. (2016). Towards smart factory for industry 4.0: a self-organized multi-agent system with big data based feedback and coordination. Computer Networks, 101, 158–168. https://doi.org/https://doi.org/10.1016/j.comnet.2015.12.017.

Zawadzki P., and Zywicki K. (2016) Smart Product Design and Production Control for Effective Mass Customization in the Industry 4.0 Concept. *Management and Production Engineering Review*, 7, pp. 102–105.

Zhang, M., and Awasthi, A. (2014). Using Six Sigma to achieve sustainable manufacturing. Proceedings of the 2014 International Conference on Innovative Design and Manufacturing (ICIDM), 311–317. https://doi.org/10.1109/IDAM.2014.6912713

8 Machine Learning Perspective in Additive Manufacturing

Mayuresh S. Suroshe[1], Vaibhav S. Narwane[1], and Rakesh D. Raut[2]

[1]Department of Mechanical Engineering, K. J. Somaiya College of Engineering, Mumbai, India

[2]Department of Operations and Supply Chain Management, National Institute of Industrial Engineering (NITIE), Mumbai, India

8.1 INTRODUCTION

To increase the quality and sustainability of industrial activities, smart manufacturing systems with innovative solutions are required. As such, AI-driven tools influenced by enabling technologies of Industrial 4.0 are equipped to create new industrial paradigms. Therefore, ML, a branch of AI, has emerged as the main focus area for today's industrial giants (Cioffi et al., 2020). ML is a system with the ability to learn from the data (Baumann et al., 2018). It is a process of teaching the computer to make predictions, find patterns, and make decisions with minimal human intervention. It creates models with the help of programs that use past results and experience to predict future outcomes (Chopra and Priyadarshi, 2019). Some commonly used ML algorithms are Supervised learning, Unsupervised learning, and Reinforcement learning (Alabi et al., 2018). Product recommendations, data mining, image and speech recognition, autonomous cars, fraud detection, etc. are some of the common applications of ML (Chopra and Priyadarshi, 2019). Nowadays, ML algorithms are effectively implemented in different industries to optimize operations such as supply chain forecasting, surface roughness, tool wear, scheduling of machines, and so forth (Chopra and Priyadarshi, 2019). Technological advancements in ML are occurring rapidly. As such, it can also process large amounts of data, including audios and videos that can help predict anomalies to prevent future breakdowns in an aircraft. It also helps to determine a specific sound of an aircraft engine operating correctly under the quality test (Web reference 1).

Additive Manufacturing (AM) or 3D printing is a process where 3D designed data, such as 3D CAD design, are used to construct 3D parts by adding layers of materials (Hu and Mahadevan, 2017; Cho et al., 2000). AM has attracted attention across the world because it solves complex designs, without assembly and with no waste. Therefore, it is preferred over conventional techniques when mass productions and fast prototypes are required. Previous studies on AM proved that it is tough to find a relationship between the dimensional parameter and process parameter by using conventional methods in AM (Yang and Zhang, 2017). The use of ML helps enhance the quality of FDM parts by enabling optimum choice of process parameters as it is challenging to get improved results using conventional techniques (Yang and Zhang, 2017).

The study tries to address the following Research Objectives (ROs) in the context of applications of ML in several areas of manufacturing. This study also aims to understand the current trends in the domain of ML in manufacturing and the significant challenges that may arise while applying ML techniques in the AM processes.

RO1: Understanding the different applications of ML in manufacturing.
RO2: Understanding the issues and challenges of ML in AM.
RO3: Proposing a framework of ML for AM.

The rest of the paper is structured as follows: Section 8.2 provides a brief about related works on the application of ML in manufacturing and AM. The proposed framework is presented in section 8.3. A case study implementation is detailed in section 8.4. Section 8.5 presents the results and discussion. Finally, section 8.6 concludes with future directions and limitations of the study.

8.2 LITERATURE SURVEY

This section gives a review of existing and related work on ML in manufacturing and AM. It also presents several applications where ML can be employed to improve the performance of AM operations. The research methodology adopted for the literature survey is illustrated in Figure 8.1.

ML techniques are often used in industries to enhance organizational efficiency as ML tools and machines collaborate to reduce mistakes and learn from previous mistakes to avoid the same in the future. Several ML tools can be employed to predict the deviation in surface roughness in real-time (Pimenov et al., 2018). Additionally, ML tools can be used to predict and monitor operations in the manufacturing sector (Chopra and Priyadarshi, 2019; Mohamed et al., 2015). The HVAC systems used in large buildings or offices are often inefficient because they do not focus on variables such as energy costs, weather patterns, and thermal

FIGURE 8.1 Methodology for literature survey.

properties of the building, etc. These issues can be eliminated using cloud-based software that use ML to solve these issues. ML segregates data and helps find factors like gas, electric, solar, and steam power important to the cooling and heating process. This analysis helps optimize the energy usage such that it results in a reduction of 10–25% of the overall energy consumption. Effective use of ML and deep learning enables advanced analytics and offers highly optimized operations in the manufacturing sector. ML tools offer benefits such as reducing costs, coping with variable demand, higher productivity, reducing downtime, etc. (Wang et al., 2018). ML could be influential in improving robot-like structures as it offers the ability to track different types of objects using cameras. Autonomous vehicles also work similarly, bringing visual information in the form of pixels and storing them in ML tools that help to guide the autonomous vehicle.

The following table provides a summary of the papers published in the field of ML for manufacturing. Table 8.1 highlights detailed research work carried out in various economies such as the USA, France, Russia, Japan, Austria, Malaysia, China, Singapore, South Africa, India, South Korea, etc. This demonstrates that all economies around the globe are desperate to adopt ML in various fields like ML in manufacturing. Table 8.1 also illustrates several aspects of papers such as publications, types of paper, tools, and techniques used.

To detect defects in AM parts using ML tools, Park et al. (2016) explained the method they employed to detect wear, scratches, burns, and dirt on the surface of the part. For surface defect detection, the authors have tested several networks to build an appropriate structure. They have used CNN with image analysis to identify defects in the targeted area of the image. Image sensors are used to find different types of defects at distinct locations in the real image. This method is beneficial, as it saves time and cost by enhancing overall performance. Gobert et al. (2018) described the application of the SVM technique to capture defects or discontinuities in the metal powder bed fusion AM process by enabling in-situ sensor monitoring. During the build process, a high-resolution DSLR was used to collect multiple images at each build layer. The linear SVM evaluates and extracts layer-wise image stacks and multi-dimensional features. The data obtained from the 3D CT Scan helps determine the porosity, inclusions, incomplete fusion, and other flaws with the exact location that is needed to train classifiers. Stanisavljevic et al. (2020) aimed to detect the interferences in the AM process. Initially, data collected from the sensors of a 3D printer enable the detection of interferences (vibration). Then, the pre-processed data is compared with the transformed data. Finally, the DPS predicts interferences accurately from the data collected by the sensors. Alabi et al. (2018) explored the significance of big data and ML techniques for several applications in the domain of AM. The authors also explained how ML techniques can be employed to detect defects during the build process in 3D printing. Pimenov et al. (2018) explained how the surface roughness deviation in real-time can be predicted by considering tool wear, flank wear, and drive power. Their study ensures the effective use of CNC machine for real-time monitoring of cutting power (N). It also estimates machining time, tool wear, and cutting power that enables the assessment of surface roughness deviation while using the cutting tool efficiently. The various AI tools used in this study are RFs, Regression trees, MLP, and Radial based

TABLE 8.1
Research Papers of ML in Manufacturing Domain

Serial No.	Publications	Tools & Techniques Used	Type of Paper
1	Alabi et al. (2018)	Artificial Intelligence, Additive Manufacturing, Machine Learning Techniques.	Theoretical paper study
2	Yang and Zhang (2017)	Neural network, Fused deposition modelling, Genetic algorithm.	Simulation
3	DeCost et al. (2017)	Additive manufacturing, machine learning	Conceptual paper framework
4	Gobert et al. (2018)	Support vector machine, Powder bed fusion, Process monitoring, Machine learning techniques.	Others
5	Khanzadeh et al. (2018)	Porosity prediction, Additive manufacturing, Supervised learning.	Others
6	Sood et al. (2012)	Thermoplastic, Artificial neural network, Quantum behaved particle swarm optimization algorithm, Fused deposition modelling, Resilient back propagation.	Literature review paper
7	Qi et al. (2019)	Machine learning, 3D printing, Additive manufacturing, Neural network.	Others
8	Baumann et al. (2018)	Additive manufacturing, survey, artificial neural network, machine learning.	Literature review paper
9	Stanisavljevic et al. (2020)	Additive manufacturing, 3D-printer, data processing system, interference detection, feature engineering, machine learning.	Others
10	Hu and Mahadevan (2017)	Additive manufacturing, Uncertainty quantification, Uncertainty management.	Literature review paper
11	Yao et al. (2017)	Hybrid Machine Learning Algorithm, Hierarchical Clustering of AM Design Features, machine learning.	Conceptual paper framework
12	Abadi et al. (2016)	Machine learning, Tensor Flow System, Tensor Flow's Python API.	Others
13	Arzi and Herbon (2000)	Flexible manufacturing system, Distributed Production Control System.	Simulation
14	Chopra and Priyadarshi (2019)	Machine learning, Industry 4.0, Manufacturing, Automation	Theoretical paper study
15	Park et al. (2016)	Deep neural networks, Machine vision, Defect detection.	Others
16	Pimenov et al. (2018)	Wear, Cutting power, Random forest, Face milling, Surface roughness.	Literature review paper

TABLE 8.1 (Continued)
Research Papers of ML in Manufacturing Domain

Serial No.	Publications	Tools & Techniques Used	Type of Paper
17	Shang and You (2019)	Big data, Machine learning, Smart manufacturing, Process systems engineering.	Literature review paper
18	Shigaki and Narazaki (1999)	Knowledge acquisition, machine learning, product innovation, Neural network.	Conceptual paper framework
19	Wang et al. (2018)	Deep learning, Data analytics, Smart manufacturing, Computational intelligence.	Theoretical paper study
20	Wu et al. (2017)	Support vector machines, Predictive modelling, prognostics and health management, random forests, artificial neural networks, Tool wear prediction.	Literature review paper
21	Yang et al. (2018)	Laser micromachining, Laser drilling, Acousto-optic deflector, Machine learning, ANN.	Others
22	Yin et al. (2020)	Deep neural networks, Machine learning, Parallel systems, Smart dispatch.	Conceptual paper framework

functions. RFs offer the highest accuracy among all AI tools. Khanzadeh et al. (2018) explained how supervised ML is used to predict porosity in an AM product. The main aim of their study was to identify the relation between defect occurrence and melt pool characteristic in AM products. A prediction method uses the morphological characteristics of a melt pool boundary. The outcome of experiments shows that the K-Nearest Neighbour (KNN) algorithm offers the highest accuracy (98.44%) among various other classifiers.

To improve and optimize AM processes using ML tools, Baumann et al. (2018) explained the application of ML and its use to improve various parameters like process monitoring and control as well as to improve the quality of manufactured AM products. The study is focused on ML in 3D printing for various applications. Qi et al. (2019) discussed how ML can be applied in AM processes from design to post-processes. Their study explained the progress of implementing NNs to several aspects of AM processes, like in-situ monitoring, material selection, quality, model design, manufacturing, etc. They also explained the current achievements of researchers in applying ML in AM. Hu and Mahadevan (2017) explored how Uncertainty Quantification (UQ) and Uncertainty Management (UM) can overcome the variation in the quality of AM processes. They also described the current scenario of research in UQ and UM and explained how these techniques can be employed effectively to improve the quality of AM products. DeCost et al. (2017)

explained how ML techniques and computer vision systems are used to find the powdered raw material used for AM products. The description algorithm and feature detection techniques are used to create micrographs. The powder micrographs (images) can then be analysed to classify the commercial powder feedstock material. Their system correctly identifies the material system with the help of powder images with an accuracy greater than 95%.

The performance parameters of numerous AM processes influence the subsequent part quality to varying degrees. An inappropriate choice of process parameters will lead to lower quality of the AM part. Therefore, the selection of optimal process parameters can be a challenging task for an expert machine operator. To analyse the influence of process parameters on AM part, Sood et al. (2012) explained the effect of various process parameters such as raster angle, orientation, layer thickness, air gap, and raster width on sliding wear of the product manufactured by the FDM process. The authors used microphotographs to explain the mechanism of wear. The optimization algorithm named QPSO helps to obtain get optimal parameters. The complexity of the FDM process and the nonlinear response of its parameters encourages the use of ANN as it offers more accuracy with fewer data and efficiently maps the input-output relationship. Yang and Zhang (2017) developed a Genetic Algorithm-Backpropagation (GA-BP) model that can optimize process parameters in the FDM machine. The results show that it is successfully implemented in process parameter prediction and can be used to solve many other complex problems. The GA-BP model predicts more accurate results than the BP model. Yao et al. (2017) explained the hybrid ML technique or hierarchical clustering that provides feasible designs by recommending design features for inexperienced designers and by targeting components based on similarities, which result in the dendrogram. The previous data (trained classifier) is utilized to train a classifier (SVM) to determine the final sub-cluster that contains recommended features. Wu et al. (2017) collected sample datasets of 315 milling tests. The authors used ML algorithms such as ANN, SVM, and RF and compared their outputs. The output stated that RFs have shown the most accurate results by considering performance parameters such as training time, R-Squared, and MSE.

To overview the recent developments of ML in the manufacturing sector, Chopra and Priyadarshi (2019) explained several algorithms and techniques involved in ML and their impact on the manufacturing sector. The authors also discussed future directions and challenges to face while employing ML in manufacturing. Yang et al. (2018) stated that ANN is the most accurate and beneficial ML method among other ML techniques, such as SVM, kNN, and Decision tree, that are employed to predict the result of the laser drilling process. Shigaki and Narazaki (1999) proposed a sintering process in steel and iron making plant using a multi-layered neural network approach. To obtain a sinter with desirable properties, a certain amount of chemical composition is required. An operational condition gives the exact amount of chemical proportion and heat input required to get processed output with the appropriate specifications. Abadi et al. (2016) explained the use of the Tensorflow platform in large-scale systems for experimenting on various operations. It is an effective platform in which various computational resources are available for training models. The authors explained the extensive use of Tensorflow and its accuracy during implementation. Arzi and Herbon (2000) explained how the effective use of an adaptive control system could improve the performance of the production system. The experiment carried out shows the difference in the performance of the production system with the use of the Distribution Production Control System and without the use of the Distribution Production Control System in random environments. Shang and You (2019) focused on the recent developments in ML and data analytics. The authors also focused on the control, monitor, and optimization of manufacturing processes as well as analyzing the gap between the current research status and actual requirement in the future. Wang et al. (2018) explained the computational methods used in deep learning and their application to make manufacturing smarter and to improve the system performance in manufacturing. The authors also discussed various models of deep learning. Yin et al. (2020) explored the challenges and characteristics of the new generation AI dispatch system in power systems. The study also explained various applications of ML algorithms in smart generation control, optimal use of power, security assessment, and smart dispatch. Finally, the study analyses the feasibility of employing robotics in dispatching system as well as the challenges that may emerge during its implementation.

8.3 PROPOSED FRAMEWORK

Usually, production lines with several printing machines are fully connected by a relentless stream of information via multiple communication protocols to communicate with one another to offer functions. Various types of sensors could be employed to capture significant characteristics of the 3D printed object and machine parts from distinct viewpoints. These sensors measure specific parameters of each layer of the 3D printed object. The communication protocols or network gateways allow data to flow from sensors mounted on the machine to cloud storage. Generally, the data is stored, processed, analyzed, and managed by a service provider with the help of cloud storage. Suitable ML algorithms can be applied to the collected data to make future predictions and make optimized decisions with minimum human intervention to generate an accurate AM product. The proposed framework enables the generation of insights as well as recommends corrective action to be taken with the help of various ML and data analysis tools. As a result, this proposed framework could be beneficial for industrial experts. The proposed framework is illustrated in Figure 8.2. It explains the details of the implementation plan from data collection to the resultant optimized data in the AM process.

8.4 CASE STUDY IMPLEMENTATION

For this study, five distinct parameters and their levels were taken into consideration. The process parameters, symbols, and their levels are illustrated in Table 8.2.

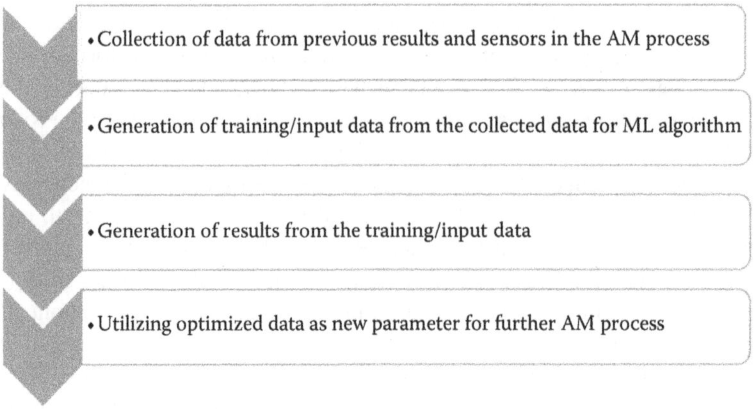

- Collection of data from previous results and sensors in the AM process

- Generation of training/input data from the collected data for ML algorithm

- Generation of results from the training/input data

- Utilizing optimized data as new parameter for further AM process

FIGURE 8.2   Implementation of ML in additive manufacturing (Baumann et al., 2018).

TABLE 8.2
Process Parameters, Symbols, and Their Levels (Yang and Zhang, 2017)

Process parameter	Symbols	Levels of Process Parameters			
		L1	L2	L3	L4
Cable width offset in mm	P	0.3	0.25	0.2	0.15
Layer thickness in mm	Q	0.25	0.2	0.15	0.1
Filling speed in mm/s	R	60	50	40	30
Extrusion speed in mm/s	S	40	35	30	25
Fallback speed in mm/s	T	75	60	45	30

8.4.1 ACQUISITION OF EXPERIMENTAL SAMPLES

The standard part of "Letter-H" geometry is employed to evaluate the accuracy of the printed parts (Yang and Zhang, 2017). The dimensional parameters that are to be measured are p, q, r, s, and t. In the standard part, p, q and r, s correspond to the x and y directions, respectively whereas, the t parameter corresponds to the z-direction. The experimental samples can appear in Table 8.3 (Annexure I). Figure 8.3 shows the standard part by FDM.

By using an FDM printing machine, the parts are manufactured based on 32 sets of process parameters tabulated in Table 8.3. The printed parts can be measured to get experimental samples. The difference between actual and theoretical size will give dimensional errors (Yang and Zhang, 2017). In this study, the dimensional errors are denoted as Δp, Δq, Δr, Δs, Δt. The experimental samples are given in Table 8.3. The process parameters (P, Q, R, S, T) are considered as the "input dataset" and dimensional errors (Δp, Δq, Δr, Δs, Δt) as the "target dataset" for simulation in MATLAB neural network Toolbox. Variable 1 is the "input dataset",

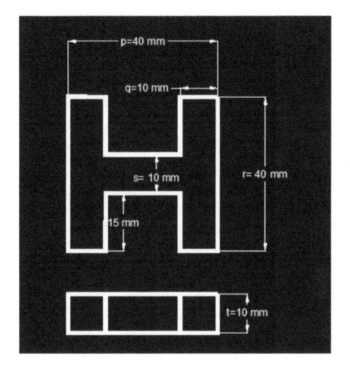

FIGURE 8.3   Standard part by FDM (Yang and Zhang, 2017).

which is a 5×32 matrix, representing the 32 samples of 5 elements. Similarly, variable 2 is the "target dataset", which is a 5×32 matrix, representing the 32 samples of 5 elements after transpose of datasets.

A Neural Network is a computational system designed with layers of connected nodes to work like the human brain. It is primarily used to overcome the complex problems occurring in day to day life. It learns from data and can be trained to predict patterns, anomaly detection, and forecast trends, etc. MATLAB Neural Network Toolbox offers a powerful platform to implement and design network models. It is a simple and user-friendly platform available for beginners. The Neural Network Toolbox can also be used to create, train, evaluate, visualize, and simulate network. The performance of NNs can be analyzed using mean square error and regression analysis.

A Neural Network Toolbox is a tool that provides an insight into how a dataset with numeric input maps the target datasets. It breaks down sample training sets into the input and target dataset. The datasets can be trained over many training samples to get the desired output. The behaviour of the neural network can be visualized by the layers and weights of those connections. The specified learning rule helps improve the performance of the NN. During training, it automatically updates weights and bias of the NN until the local minimum in the error function or desired output is reached. The simulation can also be performed using classification and regression algorithms on images, text, and other types of data by using networks such as Convolutional Neural Network and Long Short Term Memory. By using the deep network designer application, these networks can design, analyze, and train graphically. The experiment manager application can be used to examine codes and results, compare codes from different experiments, and manage deep learning experiments. It can also be employed to monitor, train, and visualize layer activations.

8.5 RESULTS AND DISCUSSION

8.5.1 NEURAL NETWORK ARCHITECTURE

In this case study, the network consists of three layers viz. the input layer, hidden layer, and output layer. The most challenging part of network architecture is the selection of the number of hidden layers and the number of neurons in each layer (Yang and Zhang, 2017). For this simulation, a single hidden layer with ten neurons was selected to reduce the training error. Thus, the structure of the neural network for this model is 5–10–5. Figure 8.4 shows the structure of the neural network.

8.5.2 CHOOSING OF THE TRAINING ALGORITHM

The network is trained to fit the input and target datasets. In this study, the Levenberg-Marquardt backpropagation algorithm is employed as it is the fastest algorithm in the toolbox. The Levenberg-Marquardt backpropagation algorithm (trainlm) typically requires more memory but less time and hence it is widely used in NN applications. In this study, the number of samples is only 32. As a result, the supervised algorithm is preferred. While training, if performance measures stop improving, then training will stop, else it may increase the MSE of validation samples. If similar samples are trained multiple times, it offers different results each time because of distinct initial conditions such as weight and bias, etc. The scaled conjugate gradient backpropagation (trainscg) algorithm may be preferred when there is not enough memory to use the "trainlm" algorithm.

The neural network is trained to reduce output errors and optimize values of weight and bias. A training function "trainlm" is taken as it automatically updates weight and bias values to offer optimized output. The selection of hidden and output

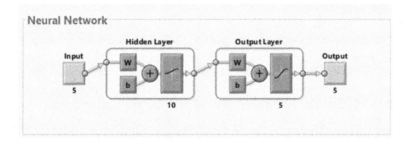

FIGURE 8.4 Structure of Neural Network.

FIGURE 8.5 Network configuration.

layer functions has a massive impact on the prediction accuracy of the neural network (Yang and Zhang, 2017). For this simulation, "LOGSIG" is used as the transfer function of hidden and output layer neurons. Whereas the adaption learning function and performance function used are "LEARNGDM" and "MSE", respectively. After the identification of network structure and parameters, the next step is the training of the network. The network type chosen is "Feed-forward back-propagation" as illustrated in the above network configuration window (Figure 8.5).

For the simulation in MATLAB neural network toolbox, the process parameters (P, Q, R, S, T) are considered as input dataset and dimensional errors (Δp, Δq, Δr, Δs, Δt) as target dataset. The weight and bias are chosen by the training network randomly to optimize using the "trainlm" training function. The training progress can be checked in the training window as progress is timely updated. Validation checks and gradient magnitude are the two most important factors that are to be considered while training. The gradient has a very small value as it converges to a minimum of performance (global minima). If the gradient value is less than 1e-5, then the training stops.

The progress of the model after training can be validated as shown in Figure 8.6. It gives a progress report of parameters such as the number of iterations, validation checks, momentum parameter value, time taken for iterations, and gradient value. To observe the performance of the model, it generates some plots such as Performance plot, Training state plot, and Regression plot.

An epoch indicates the total number of cycles that the training dataset has completed with assumed weights and biases. 15 iterations mean that the input is

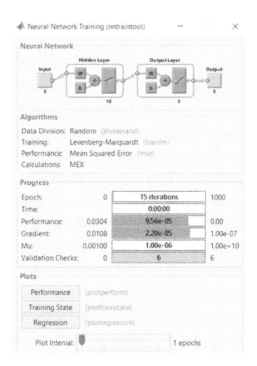

FIGURE 8.6   Validation of model after training.

trained 15 times and compared with 32 samples that are to be trained. It auto-matically stops training when the error rate increases continuously. "Mu" (momentum constant/parameter) is the adaption parameter that is used in the Levenberg-Marquardt optimization process while calculating the updated para-meters as well as to train the neural network. It tells about the amount of change in bias and weights after every iteration. It helps to speed up gradient descent as the value of "Mu" affects the error of convergence. The "Mu" value generally lies between 0 and 1. The network may sometimes get stuck to the local minimum and fail to converge. In that scenario, the momentum constant helps to avoid the problem of local minimum validation checks. The validation check shows the number of successive iterations in which the validation performance fails to decrease. Mu is the adaptation parameter that generalizes the non-trained data. The training will stop if the validation error rate successively increases for more than six epochs.

In the regression plot, R values measure the correlation between outputs and target value. An R-value close to 1 means there is very little error between output and target value or close relationship as well as R-value near 0 means high error between output and target value or random relationship. The regression plots are displayed in Figure 8.7. In the plot, the dashed line signifies target values. The outcome shows that the R-value is large and close to 1, which indicates that the training data is a good fit.

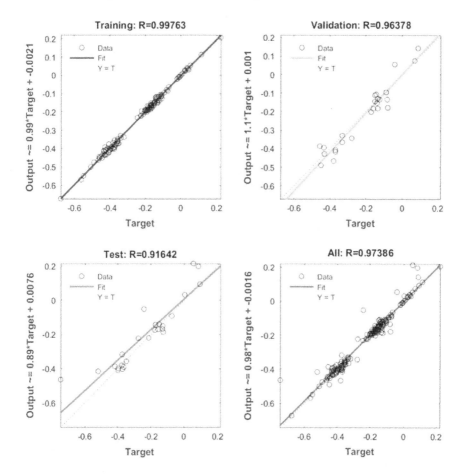

FIGURE 8.7 Regression plot.

Mean Square Error (MSE) is an average of the squares of the error between target and output values. The smaller the MSE, the closer the results are to the best fit or results with low error. Thus, a well-trained NN should have very low MSE (close to zero). In the performance plot, the best validation performance is 0.002237 at epoch 9. Out of 15 iterations of the system to classify or compare the data, the best performance is obtained at the ninth iteration. After training, the error reduces for more epochs but can start to increase instead when the network overfits the data. If validation error increases six consecutive times, then the training will stop. It takes the best training performances among all epochs with the least error. For this example, the plot indicates that the test and validation curves are not far from each other hence it signifies that the performance plot does not point out any serious problems with the training. In case some overfitting is present then the test curve will increase significantly before the validation curve. These signs indicate the presence of random error in the data. Refer Figure 8.8 for the performance plot.

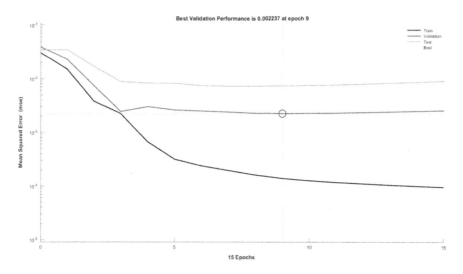

FIGURE 8.8   Performance plot.

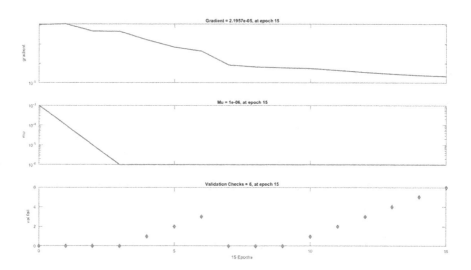

FIGURE 8.9   Training state plot.

In the training state plot, once the samples are trained, the checking of validation samples is necessary to ensure the performance of the model. The validation samples are samples of input targets that offer an impartial estimation of a model fit on training datasets. The outcome shows that there are 6 validation checks and that the model performs better at epoch 15. The gradient is the value of backpropagation on each iteration in the logarithmic scale. It shows that the bottom of the local minimum of the target function has been reached. Refer Figure 8.9 for the training state plot.

8.6 CONCLUSION

Companies are pushing to discover new techniques of creating sustainable industrial value because of the rise of smart technologies, the changes in customer demand, as well as sustainable economic development of manufacturing industries. The industries are already taking initiatives to integrate ML approaches with AM for improving agility. However, because of the inadequate research regarding the adoption of ML technology in the AM, industries are facing several problems while adopting ML technology in the AM. Therefore, the aim of this research was to identify the prevailing studies carried out in the domain of ML in manufacturing and the significant challenges that emerge while adopting ML technology in the AM.

This study presents an overview of research carried out in the domain of ML in the manufacturing sector and identifies the challenges of adopting ML in the AM. The study mainly focuses on the applications where ML algorithms are widely utilized in the manufacturing sector to improve the workflow, quality of the printed part, anomaly detection, operational efficiency, and supply chain management, etc. The study also discusses the current challenges in adopting ML in AM. The proposed framework explains how the results of AM operation can be enhanced by using ML algorithms. The case study focuses on validating the results of a simulation to evaluate the dimensional accuracy of FDM parts by optimizing the process parameters using a neural network toolbox in MATLAB. In this study, five process parameters (P, Q, R, S, and T) are considered and analyzed on the standard part of "Letter-H" geometry and the rest controlled at their fixed level. The network is trained to fit the input data and target data using the Levenberg–Marquardt backpropagation algorithm. The study presents the successful implementation of the feed-forward BP model in the FDM process. The simulation results of the performance plot, regression plot, and training state plot shows that the model has sufficient accuracy. Thus it can be concluded that optimizing process parameters, operational efficiency, and accuracy of the AM parts can be enhanced and analyzed.

The study encourages practitioners and researchers to explore more areas where ML can be employed to improve operations, quality, and productivity in their organization. It also offers detailed a description of the Neural Network Toolbox in MATLAB and its features. The outcome of the model can be improved by considering the effect of process parameters on the physical properties of the part. Also, this study only analyses five operating parameters. Therefore, the various other process parameters, such as layer thickness, raster angle, and width, air gap, etc., can be considered for future studies. Further research must be conducted on the merits and demerits of using both techniques together with experimental justification, specifically in the adoption of ML technology in the AM. In the future, several new challenges may hinder the adoption of ML technology in the AM. In such a case, further action plans can be identified through literature and expert opinions to overcome newly identified challenges. Lastly, these technologies also deliver social, economic, and environmental benefits and opportunities. The combination of these technologies conforms with the concept of sustainability. The manufacturers must follow all three scopes simultaneously to create sustainable industrial value. Thus, the study recommends superiors to study Industry 4.0 enabling technologies systematically for long-term sustainability. ML, therefore, can offer several opportunities for improving agility in the future.

REFERENCES

Abadi, M., Agarwal, A., Barham, P., Brevdo, E., Chen, Z., Citro, C., Corrado, G., Davis, A., Dean, J., Devin, M., Ghemawat, S., Goodfellow, I., Harp, A., Irving, G., Isard, M., Jia, Y., Jozefowicz, R., Kaiser, L., Kudlur, M., Levenberg, J., Mane, D., Monga, R. Moore, S., Murray, D., Olah, C., Schuster, M., Shlens, J., Steiner, B., Sutskever, I., Talwar, K., Tucker, P., Vanhoucke, V., Vasudevan, V., Viegas, F., Vinyals, O., Warden, P., Wattenberg, M., Wicke, M., Yu, Y., and Zheng, X. (2016). Tensorflow: Large-scale machine learning on heterogeneous distributed systems. *arXiv preprint arXiv:1603.04467*.

Alabi, M.O., Nixon, K., and Botef, I. (2018). A survey on recent applications of machine learning with big data in additive manufacturing industry. *American Journal of Engineering and Applied Sciences*, 11(3), pp. 1114–1124.

Arzi, Y., and Herbon, A. (2000). Machine learning based adaptive production control for a multi-cell flexible manufacturing system operating in a random environment. *International Journal of Production Research*, 38(1), pp. 161–185.

Baumann, F.W., Sekulla, A., Hassler, M., Himpel, B., and Pfeil, M. (2018). Trends of machine learning in additive manufacturing. *International Journal of Rapid Manufacturing*, 7(4), pp. 310–336.

Cho, H.S., Park, W.S., Choi, B.W., and Leu, M.C. (2000). Determining optimal parameters for stereolithography processes via genetic algorithm. *Journal of Manufacturing Systems*, 19(1), pp. 18–27.

Chopra, V., and Priyadarshi, D. (2019). Role of machine learning in manufacturing sector. *International Journal of Recent Technology and Engineering*, 8(4), pp. 2320–2328.

Cioffi, R., Travaglioni, M., Piscitelli, G., Petrillo, A., and De Felice, F. (2020). Artificial intelligence and machine learning applications in smart production: Progress, trends, and directions. *Sustainability*, 12(2), Art. no. 492.

DeCost, B.L., Jain, H., Rollett, A.D., and Holm, E.A. (2017). Computer vision and machine learning for autonomous characterization of AM powder feedstocks. *Jom*, 69(3), pp. 456–465.

Gobert, C., Reutzel, E.W., Petrich, J., Nassar, A.R., and Phoha, S. (2018). Application of supervised machine learning for defect detection during metallic powder bed fusion additive manufacturing using high resolution imaging. *Additive Manufacturing*, 21, pp. 517–528.

Hu, Z., and Mahadevan, S. (2017). Uncertainty quantification and management in additive manufacturing: current status, needs, and opportunities. *The International Journal of Advanced Manufacturing Technology*, 93(5–8), pp. 2855–2874.

Khanzadeh, M., Chowdhury, S., Marufuzzaman, M., Tschopp, M.A., and Bian, L. (2018). Porosity prediction: Supervised-learning of thermal history for direct laser deposition. *Journal of Manufacturing Systems*, 47, pp. 69–82.

Mohamed, O.A., Masood, S.H., and Bhowmik, J.L. (2015). Optimization of fused deposition modeling process parameters: A review of current research and future prospects. *Advances in Manufacturing*, 3(1), pp. 42–53.

Park, J.K., Kwon, B.K., Park, J.H., and Kang, D.J. (2016). Machine learning-based imaging system for surface defect inspection. *International Journal of Precision Engineering and Manufacturing-Green Technology*, 3(3), pp. 303–310.

Pimenov, D.Y., Bustillo, A., and Mikolajczyk, T. (2018). Artificial intelligence for automatic prediction of required surface roughness by monitoring wear on face mill teeth. *Journal of Intelligent Manufacturing*, 29(5), pp. 1045–1061.

Qi, X., Chen, G., Li, Y., Cheng, X., and Li, C. (2019). Applying neural-network-based machine learning to additive manufacturing: Current applications, challenges, and future perspectives. *Engineering*, 5(4), pp. 721–729.

Shang, C., and You, F. (2019). Data analytics and machine learning for smart process manufacturing: Recent advances and perspectives in the big data era. *Engineering*, 5(6), pp. 1010–1016.

Shigaki, I., and Narazaki, H. (1999). A machine-learning approach for a sintering process using a neural network. *Production Planning & Control*, 10(8), pp. 727–734.

Sood, A.K., Equbal, A., Toppo, V., Ohdar, R.K., and Mahapatra, S.S. (2012). An investigation on sliding wear of FDM built parts. *CIRP Journal of Manufacturing Science and Technology*, 5(1), pp. 48–54.

Stanisavljevic, D., Cemernek, D., Gursch, H., Urak, G., and Lechner, G. (2020). Detection of interferences in an additive manufacturing process: An experimental study integrating methods of feature selection and machine learning. *International Journal of Production Research*, 58(9), pp. 2862–2884.

Wang, J., Ma, Y., Zhang, L., Gao, R. X., and Wu, D. (2018). Deep learning for smart manufacturing: Methods and applications. *Journal of Manufacturing Systems*, 48, pp. 144–156.

Web references 1: https://www.forbes.com/sites/louiscolumbus/2019/08/11/10-ways-machine-learning-is-revolutionizing-manufacturing-in-2019/#73079d22b404 (Accessed on 25 June, 2020)

Wu, D., Jennings, C., Terpenny, J., Gao, R. X., and Kumara, S. (2017). A comparative study on machine learning algorithms for smart manufacturing: tool wear prediction using random forests. *Journal of Manufacturing Science and Engineering*, 139(7).

Yang, C., Hu, H., and Zhang, H. (2018). Modeling AOD-driven laser microvia drilling with machine learning approaches. *Journal of Manufacturing Processes*, 34, pp. 555–565.

Yang, H.D., and Zhang, S. (2017). Precision prediction model in FDM by the combination of genetic algorithm and BP neural network algorithm. *Journal of Measurements in Engineering*, 5(3), pp. 134–141.

Yao, X., Moon, S.K., and Bi, G. (2017). A hybrid machine learning approach for additive manufacturing design feature recommendation. *Rapid Prototyping Journal*, 23(6), pp. 983–997.

Yin, L., Gao, Q., Zhao, L., Zhang, B., Wang, T., Li, S., and Liu, H. (2020). A review of machine learning for new generation smart dispatch in power systems. *Engineering Applications of Artificial Intelligence*, 88, Art. no. 103372.

ANNEXURE

ANNEXURE I TABLE 8.3
Experimental Samples (Yang and Zhang, 2017)

Serial	P	Q	R	S	T	Δp	Δq	Δr	Δs	Δt
1	0.3	0.25	60	55	75	−0.676	−0.185	−0.49	−0.179	−0.021
2	0.3	0.25	30	35	60	−0.375	−0.141	−0.374	−0.073	−0.202
3	0.3	0.2	60	25	75	−0.449	−0.151	−0.405	−0.145	0.113

(Continued)

ANNEXURE I TABLE 8.3 (Continued)
Experimental Samples (Yang and Zhang, 2017)

Serial	P	Q	R	S	T	Δp	Δq	Δr	Δs	Δt
4	0.3	0.2	30	45	30	−0.363	−0.15	−0.336	−0.128	−0.109
5	0.3	0.15	50	55	60	−0.402	−0.171	−0.373	−0.168	−0.15
6	0.3	0.15	40	35	45	−0.423	−0.201	−0.355	−0.155	−0.115
7	0.3	0.1	50	25	75	−0.436	−0.189	−0.384	−0.244	−0.079
8	0.3	0.1	40	45	30	−0.444	−0.224	−0.421	−0.224	0.01
9	0.25	0.25	60	45	60	−0.562	−0.178	−0.459	−0.167	0.221
10	0.25	0.25	30	25	45	−0.43	−0.134	−0.332	−0.089	−0.08
11	0.25	0.2	60	35	75	−0.52	−0.157	−0.398	−0.151	0.053
12	0.25	0.2	30	55	30	−0.428	−0.167	−0.379	−0.13	0.034
13	0.25	0.15	50	45	60	−0.46	−0.168	−0.349	−0.105	0.022
14	0.25	0.15	40	25	45	−0.417	−0.193	−0.381	−0.164	0.043
15	0.25	0.1	50	35	75	−0.406	−0.22	−0.406	−0.211	0.001
16	0.25	0.1	40	55	30	−0.746	−0.214	−0.399	−0.277	0.002
17	0.2	0.25	50	35	30	−0.55	−0.131	−0.427	−0.124	0.053
18	0.2	0.25	40	55	75	−0.508	−0.127	−0.393	−0.093	0.053
19	0.2	0.2	50	45	45	−0.448	−0.149	−0.37	−0.142	-0.042
20	0.2	0.2	40	25	60	−0.434	−0.189	−0.357	−0.169	−0.067
21	0.2	0.15	60	35	30	−0.373	−0.176	−0.349	−0.173	0.092
22	0.2	0.15	30	55	75	−0.429	−0.119	−0.434	−0.136	0.038
23	0.2	0.1	60	45	45	−0.455	−0.134	−0.279	−0.083	0.069
24	0.2	0.1	30	25	60	−0.466	−0.189	−0.355	−0.17	−0.013
25	0.15	0.25	50	25	30	−0.38	−0.138	−0.37	−0.124	0.088
26	0.15	0.25	40	45	75	−0.418	−0.129	−0.374	−0.075	0.083
27	0.15	0.2	50	55	30	−0.376	−0.131	−0.365	−0.126	−0.243
28	0.15	0.2	40	35	60	−0.399	−0.145	−0.356	−0.108	−0.021
29	0.15	0.15	60	25	30	−0.365	−0.162	−0.339	−0.185	0.024
30	0.15	0.15	30	45	75	−0.391	−0.138	−0.32	−0.106	−0.007
31	0.15	0.1	60	55	30	−0.373	−0.2	−0.412	−0.179	−0.007
32	0.15	0.1	30	35	60	−0.433	−0.17	−0.332	−0.158	−0.149

9 Sustainability Performance Measurement Methods, Indicators, and Challenges
A Review

Nagendra Kumar Sharma[1], Wen-Kuo Chen[2], Kuei-Kuei Lai[3], and Vimal Kumar[3]

[1]Department of Business Administration, Chaoyang University of Technology, Taiwan & Department of Management Studies Graphic Era University, India
[2]Department of Marketing and Logistics Management, Chaoyang University of Technology, Taiwan
[3]Department of Information Management, Chaoyang University of Technology, Taiwan

9.1 INTRODUCTION

Rapid industrialization is good for the economy, but it also leads to several environmental threats that cannot be ignored (Varma and Kalamdhad, 2017). Therefore, currently, there is much debate on the implementation of sustainable development strategies in industries (Lekan et al., 2020). Stakeholders are giving much more emphasis to developing strategies for achieving and maintaining sustainability. Most of the industries have adopted sustainable development practices as a defensive approach so that they can avoid penalties imposed by authorities for damaging the environment (Lynch et al., 2016). But many industries have picked up a cooperative approach, wherein, industries go green so that they get a competitive advantage and increase their performance in the long run (Kushwaha and Sharma, 2016). Several researchers have worked on sustainability issues, and there are many quality research publications available on this. The sustainable development initiatives in the industries ensure rapid industrialization with optimal use of the resources. The word "sustainable development" was initially introduced by the

International Union for Conservation of Nature (IUCN) in 1980. The "Brundtland Report", which was named "Our Common Futures" (WCED, 1987), defined sustainable development as development that "*meets the needs of the present generation without compromising the ability of future generations to meet their own needs*" Moldan 2012. According to Lélé (1991), *Sustainable development has become the watchword for international aid agencies, the jargon of development planners, the theme of conferences and learned papers, and the slogan of developmental and environmental activists* (Bell and Morse, 2012). Industries put more effort into various green business strategies such as green marketing, green supply chain management, reverse logistics, and other circular economy models (Sharma and Kushwaha, 2018). These green initiatives lead to sustainable development in the long term. There are several measurement parameters for measuring the sustainability of Industries. Elkington (1997) introduced the term "triple bottom line" as a sustainability concept where emphasis has been given to sustainability measures based on three significant dimensions: environment, society, and economy. Therefore, a company's sustainable performance can be measured on these triple bottom line scales in terms of social, environmental, and financial performance (Laosirihongthong et al., 2020). Figure 9.1 shows the triple bottom line model where all the three elements (environment, society, and economy) intersect one another, which connotes sustainability in subsections of all these elements of the triple bottom line. Sustainable performance of the industries is measured individually on a social, environmental, and financial scale. The sustainability assessment parameters should respond towards three significant components: integrating social systems, assessing spatial levels, and assessing long-term and short-term outlooks (Ness et al., 2007). To accomplish the goal of sustainability it is imperative to develop significant indicators for sustainable development, which further need to be measured quantitatively and qualitatively as per the company's SDGs (Moldan et al., 2012). In Figure 9.2 a basic framework of the evolution of sustainability and measurement process has been shown.

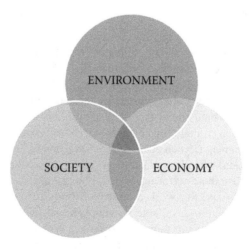

FIGURE 9.1   Triple bottom line of sustainability (Elkington, 1997).

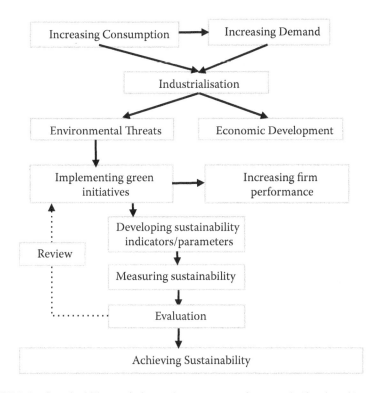

FIGURE 9.2   Sustainability evolution and measurement framework (developed by authors).

Industries are practicing green initiatives to gain sustainability. But, certain questions need to be relooked upon. These are:

- What are the methods and challenges while measuring the sustainable performance of industries?
- What are the various sustainability parameters for measurement?

Previous research has made several attempts to discuss and showcase sustainability parameters for its performance measurements and associated challenges, but there is still a lack of quality research available that shows performance parameters for sustainability at a single platform.

Therefore, the present chapter is interesting for the readers as it puts significant endeavours to present all the relevant factors and sustainability parameters altogether in one study. The study is qualitative and based on a critical review of previous studies that have been published on sustainability reporting and sustainability performance parameters. The chapter is organized into several sections, after the introduction part, the chapter has sections as a critical review of previous studies, methodology, theoretical findings and discussions, conclusion, the implication of the study and at the end future research scope.

9.2   CRITICAL REVIEW OF PREVIOUS STUDIES

In this section of the chapter, previous studies based on sustainable development parameters or indicators, measurement methods, and associated challenges are critically reviewed. The literature has been collected from "google scholar". Certain specific keywords, such as "sustainability parameters", "sustainability indicators", sustainability measurement" and "sustainability measurement challenges", were used to choose relevant articles for the study. 29 articles were selected based on the proximity to the objective of the study. The following are the results.

Hammond and World Resources Institute (1995) has focused on the various environmental indicators using a systematic approach. The study has also highlighted measurement and reporting on environmental policy performance targeting sustainable development.

KEY LEARNINGS

The study has highlighted the required characteristics of the sustainability indicators as they help quantify information, simplify information, and the indicators must be user-driven: they should be relevant to policy, and finally, should be highly aggregated. The indicators discussed in the study were suggested by OECD & UNEP: climate change, urban environmental quality, waste, water resources, soil, ocean conservation, etc. Other indicators were mentioned in the study suggested by the World Bank, which was classified as source indicators (for example agriculture, forest, water, fossil fuels, etc.) sink or pollution life-support indicators (climate change, greenhouse gases, eutrophication, etc.), life support indicators (biodiversity oceans special lands) and human-impact indicators (health, water, quality, air quality, food security, and quality, natural disaster, housing).

Gallopin (1996) has highlighted sustainability indicators and the concept of situational indicators. The study has used a system approach.

KEY LEARNINGS

The study has shown the environmental standards in form of sustainability indicators that are used for measuring sustainability performance. The study incorporates all the elements relevant to the environment, finance, and society in the performance indicators of sustainability.

Yeh and Li (1997) have conducted a study that highlights the monitoring and evaluation of rapid urban growth for sustainable development. The study is about the Pearl River Delta, China. The monitoring was done with integrated remote sensing and geographical information systems (GIS) approach.

KEY LEARNINGS

The study has found that there is a rapid change in land use. The nature of area and location based on land use are changing. There are several agricultural challenges are taking place such as the loss of fertile land for productive agriculture in the region. These land uses are of two types such as shifting from grain growing to fruit growing and shifting from agricultural to urban use. All these issues are creating socio-economic challenges and also hampering the sustainability goals of the region.

Azapagic and Perdan (2000) have worked on the measurement of the level of sustainability of the industry. The study aimed to further welcome the debate about sustainability measures.

KEY LEARNINGS

The paper proposes a basic framework with simple measurement scales for sustainability performance. The study includes indicators based on economic, social, environmental, financial, and other ethical parameters. The framework was designed in a way so that it can be a better fit in the industry. However, the study also emphasizes that there is a need for some specific indicators that are based on case to case basis. The study also highlights the Global Reporting Initiatives (GRI) for the reporting of sustainability performance and its drafting guidelines. The indicators.

Labuschagne et al. (2005) have worked on measures of sustainability based on multi-criterion decision analysis (MCDA).

KEY LEARNINGS

The available measures are not sufficient for measuring overall business sustainability and cannpt affect the operational level for developing nations and hence do not receive much attention. Global Reporting Initiatives (GRI) were introduced by the United Nations Environment Programme (UNEP) with the help of the United States, Coalition for Environmentally Responsible Economics (CERES). The goal of GRI was to enhance the quality and sincere reporting of sustainability. Two broad routes were discussed for the measurement of sustainable performance as "valuation route" and "qualitative route". The valuation route consists of financial terms such as cost-benefit analysis, net present value and in the qualitative route, MCDA is used. The study adds knowledge about the four key approaches for sustainability indicators: these are the Global Reporting Initiative (GRI), the United Nations Commission on Sustainable Development (UNCSD) Framework, the Sustainability Metrics of the Institution of Chemical Engineers (IchemE), and Wuppertal Sustainability Indicators (WSI). GRI consists of social, environmental, and economic indicators viz. labour practices, human rights, social welfare, environmental protection, economic benefits, etc. UNCSD uses indicators such as education, health, security, population control, housing, land use, ocean, freshwater conservation, economic welfare programs, consumption, and production pattern and institutional capacity, etc. IchemE also introduced sustainable parameters as resource utilization, carbon emission, waste, and in the economic head it adds profit, tax, and investment, and finally, in social criteria, it adds work culture and social cooperation. WSI contains indicators in three heads with their indicators like others such as environmental (resource utilization etc.), economic (GDP, NI, growth rate, etc.), social (health, social security, and housing, etc.), and institutional (participation, gender balance, and justice).

Hueting and Reijnders (2004) have discussed the sustainability indicators and also argued about the major parameters (social, economic, and environmental) and their inclusion in the sustainability measures.

Key Learnings

The authors argued about the inclusion of economic and social elements in sustainability indicators as these factors do not directly impact the lives and resources of future generations. Hence, both social and economic indicators should be presented separately. The study suggested that the only better indicator of sustainability is genuine savings by meeting essential conditions. There are several suggestions on the indicators based on employees and their working environment, yearly withdrawal of water resources, carbon emission, number of cars per thousand, forest, killings of human, numbers of phone lines, urban population and its sewer connections, social security. System of Environmental and Economic Accounting (SEEA) several sustainable development goals and sustainability measures.

Gallego (2006) has studied the indicators for measuring sustainable development based on economic, social, and environmental indicators.

Key Learnings

The study highlights sustainable development measures based on three criteria: economic, social, and environmental. The study has also shown that how these indicators are used by the Global Reporting Initiative (GRI). GRI is a globally accepted parameter. The indicators were classified as economic (direct economic impacts such as customers, suppliers employees, etc.), environmental (environmental materials, energy, water, emission, transport, etc.), social (labour practices, decent work, health, safety, training, and education, etc.), human rights strategy management (non-discrimination, freedom of association, child labor, disciplinary practices, etc.) and society (community, bribery and corruption, political contribution) and product responsibility (customer health, product, and services, advertising, respect, and privacy, etc.).

Hezri and Dovers (2006) have focused on sustainability indicators and their assessment policies in association with production. The study addresses two key points i.e., the utility of sustainability indicators and impact on governance.

Key Learnings

The study suggests that sustainability indicators and their significance can be verified only when there is an appropriate exploration of historical indicators and applications such as the study of public administration, environmental studies, and urban development studies. Several measures, such as biophysical accounts, eco-efficiency, and indicators set with dematerialization, state of the environment reporting, and other methodological approaches guided and reported to the Government are used for sustainability performance.

Munday and Roberts (2006) have developed and studies the approaches to monitoring and measuring sustainable development. The study has examined a selection of aggregate approaches. The paper is based on a review of previous studies.

Key Learnings

Several approaches were discussed to measure the established parameter of sustainability such as ecological footprint, environmental satellite accounts (ESAs), environmental input-output (ENVIO) tables, index of sustainable economic welfare (ISEW), and Welsh Assembly Government headline sustainability indicators. These indicators were major includes housing, crime, education, air quality, employment, wild-life, waste, and ecological footprint, etc.

Bagheri and Hjorth (2007) have conducted a study based on methodology development for the system for monitoring sustainable development and its associated practice specifically in the urban water system in Tehran.

Key Learnings

The study reveals that sustainability, in the context of water resources, is necessary which will fulfil the objective of society even today and in the future while maintaining ecological and hydrological integrity. Water sustainability can be maintained well with the help of focusing on urban water systems, sewage system stormwater, soil, air, and other socio-economic systems.

Evans et al. (2009) have studied the sustainability indicators that are useful in the performance measurement of sustainable development in the firm. The study is based on energy-based firms.

Key Learnings

The indicators were used for assessment are the effectiveness of energy conservation greenhouse gas emission during the entire life cycle, availability of natural and renewable resources, price of the electricity water consumption, land requirement, and other social impacts. The result of the study highlighted that wind power is more sustainable among other sources such as hydropower, photovoltaics, and then geothermal.

Ramos and Caeiro (2010) have developed a conceptual framework with the objective of designing and assessment of the efficiency of the sustainable development indicators.

Key Learnings

The study has proposed "good practice factors" and meta performance evaluation indicators that have been introduced for a national case study. The National

Sustainable Development Indicators System (SIDS) proposed by the Portuguese Ministry of the Environment and Land-Use Management in the year 2000 shows the effectiveness of the sustainability indicators and also aids in the overall assessment of performance scanning exercise and results. In another initiative that was started in 1997 based on the Portuguese indicators, 132 sustainable development indicators (SDIs) were proposed by the United Nations based on Pressure-State-Response (PSR) frameworks. Whereas 29 indicators were based on economic, 72 on the environment, 22 for social, and 9 were for institutional parameters. PSR has widely used frameworks for a causal chain where society and economy are considered as driving forces that put pressure on the ecosystem and it leads to change in the ecology@@@ (Huang et al., 2011).

Hermans et al. (2011) have conducted a comparative study based on four different cases based on sustainable development in the Netherlands.

KEY LEARNINGS

The results of the study found that contextual factors are more relevant than time while monitoring sustainability. Sustainable development reflects in the socio-economic and ecological structural characteristics of their region. Stakeholders have a very important role in deciding the sustainable performance of the firm.

Shen et al. (2011) have done research based on the urban sustainability indicators by using nine different practices and proposed several indicators for assessment.

KEY LEARNINGS

The main learning of the study clears that indicators are not fixed and based on the practices used by the institutions. The study has suggested comparative measures or indicators such as International Urban Sustainability Indicators List (IUSIL). This list of indicators helps in understanding the direction of drivers and the aim of each practice. The comparative analysis was based on the four key areas as economic, environmental social, and governance. Whereas environmental consists of 10 major indicators, 5 indicators were listed under economic such as consumption, production pattern, finance, etc. social criteria enlist 18 indicators for example water, education, shelter, land, and culture, etc. and finally, the governance lists 4 indicators these are participation and civic engagement, transparent, accountable and efficient governance government and sustainable management of the authorities and businesses.

Vithayasrichareon et al. (2012) have conducted a study on electricity industries and sustainable performance. The study puts efforts into evaluating the sustainability challenges in the electricity industries. The study has highlighted on 3As (accessibility, availability, and acceptability) which are used as a sustainability analytical framework.

KEY LEARNINGS

The analysis of the study shows that the main aim of sustainability in the electricity industries is to achieve the electricity demand, maintaining the security of the electricity supply, and controlling the CO2 emission while electricity generation. The parameters that can be measured under the 3A arc: Accessibility: It includes indicators such as electricity prices, electricity bills, subsidy, etc. Availability (It includes indicators like reliability, supply time, reliability operating standards, etc.) and Acceptability (it includes parameters such as strategy for nuclear power, renewable energy, greenhouse emission, and safety, etc.).

Turcu (2013) has done a study on sustainability indicators considering urban sustainability. The authors have developed an integrative set of indicators verified with the help of sustainability experts.

KEY LEARNINGS

The study highlights that the sustainability indicators are not the isolated kind of information but it has some signs of understanding the interconnections with the ecology and people. Urban sustainability indicators were classified among environmental (resources, housing, and built environment, services and facilities), economic (local jobs, access to jobs, housing prices and affordability, etc.) social (community development, safety, and income, etc.), and institutional sustainability (local authority services, community activity, and local partnership).

Yin et al. (2014) have studied the sustainability measures in 30 Chinese cities it was an empirical study. The study was about significant sustainability measures called "eco-efficiency".

KEY LEARNINGS

The term eco-efficiency can be divided into two parts "eco" which means both environmental and economic performance. To measure the eco-efficiency of the organization both financial and ecological information is required. The financial information helps in calculating numerators whereas, the ecological information works for calculating denominators. The economic indicators are the gross domestic product (GDP), net sales, and volume of products produced. The eco information can be the consumption of various natural resources such as energy, water, material, and other information related to carbon emission, material emission, and ozone depletion. The eco-efficiency is classified into three parts. The first part is a single ratio model based on economic outcomes and the ecological impact that has been highly accepted. The second part is the substitution of the numerator with other composite indicators that consist of an analysis of material flow, indicators based on energy parameters, and indicators of eco footprints. The third part uses several other models to estimate eco-efficiency such as positive matrix factorization, factor analysis, and principal components analysis. In recent scenarios, the data envelopment analysis (DEA) model has been much in practice in the analysis of eco-efficiency which has several benefits.

Feil et al. (2015) have identified and selected the indicators towards measuring sustainability in small industries particularly in furniture industries. The three key aspects as social, economic, and environmental parameters of sustainability were used to develop the indicators.

Key Learnings

The study developed the indicators as environmental indicators (solid waste, liquid waste, etc., ozone layer protection, greenhouse gases, recycling of products, reuse of products, etc.). Social indicators (employee turnover, lay-offs, employee training and development, employee education, health and safety, child labor, sexual harassment, and quality of life, etc.). Economic indicators (revenue, operating profit, net profit, tax payments wages, and operational cost, etc.)

Jurigová and Lencsésová (2015) have done a study to monitor sustainable development focusing on mountains. The study is based on a literature review. The study incorporates some specific indicators.

Key Learnings

The study mentioned the significance of the indicators proposed by the United Nations in 1992. Document Agenda 21 was set to measure sustainability. Various indicators based on socio-economic factors were set and by the United Nations by considering the societal challenges and changes affecting the business houses. These challenges and changes are significant for establishing sustainable indicators. The study highlights mainly environmental indicators, economic indicators, and social indicators considering the tourism industry.

Rickels et al. (2016) have studied the indicators for monitoring (sustainable development goals) SDGs. The study was about oceanic development in the European Union.

Key Learnings

The study has highlighted the sustainable development agenda set for 2030. It includes 17 (SDGs) including 169 targets. The 193 members of the UN General Assembly.

Agenda for Sustainable Development (2030).

Moldan et al. (2012) have worked on sustainability indicators and their development approaches. The study has highlighted various indicators measuring sustainability at all dimensions of the firm.

Key Learnings

The study emphasizes setting the target for sustainability and then measurement to get the current trend of the information. The target can be of two types known as soft targets and hard targets. The hard target is set with political influence and hence it becomes part of the significant policy. Whereas, the hard target is based on our wisdom and uses sustainable references. The indicators prepared for the measure can be well interpreted with the help of distance from the target. European Environmental Agency (EEA) in 1999 given the reference value for the sustainability measures. The indicators were divided into four groups that address key points such as the current state of the environment and human, the significance of the indicators, the status of improvement, and finally, no improvement is required at all. The reference value has been made for comparing the performance with the current and actual conditions. This measurement is basically to measure the distances between present ecological situations and the expected that is why it is also called the *"distance to target"* assessment model. The Organization for Economic Cooperation and Development (OECD) has introduced a new terminology called *"decoupling"* where a new methodology and parameters were developed to measure the decoupling of ecological pressure from economic development and growth. The OECD has listed several indicators capturing a wide spectrum of ecological issues like waste disposal, air pollution, natural resources, climate change, material use, water quality. There are another set of indicators that can be used for decoupling analysis of four areas viz. transport, manufacturing, agriculture, and energy. Other indicators were used as deforestation, ozone layer depletion, CO_2 emission, the ecological burden on pandemic and other diseases and climate change, etc.

Fazlagić and Skikiewicz (2019) have studied sustainable development measurement leading to the creative economy. The study has focused on the role of government in developing a creative economy.

KEY LEARNINGS

The study has focused on the sustainable development measure for the development of the creative economy. The measures such as transportation connectivity, education level, universities, development of small businesses, green environment, etc. are needed to be enhanced. The development of these measures will attract tourism in the region and this will also enhance the creative economy of the nation. The sustainable development measures can be improved with the help of the government and these sustainability initiatives also improve the socio-economic condition.

Albawab et al. (2020) have researched sustainability performance index with the help of **the** Multi-Criteria Decision-Making (MCDM) model and model and Stepwise Weight Assessment Ratio Analysis (SWARA)/Additive Ratio Assessment (ARAS).

KEY LEARNINGS

Five major indicators namely environmental, economical, resource-based, technological, and social were listed with 17 subsections that are: energy, CO2, greenhouse gas emissions, health and safety issues, energy density, current installed capacity, current installed capacity, capital intensities, construction, specific energy, operating cost cycle, lifetime, discharge time at rated power, specific power, cycle efficiency, growth rate and adaptability for mobile systems. The sustainability performance can be assessed by the Stepwise Weight Assessment Ratio Analysis method integrated with the Additive Ratio Assessment method (Hybrid Extended SWARA/ARAS). The evaluation process in this method can be performed with the help of three energy experts. The results of the assessment can be shown on graphical or numerical presentations.

Aksoy et al. (2020) have conducted an empirical study based on sustainability performance. The study confirms that foreign and institutional ownership has a positive impact on sustainability performance.

KEY LEARNINGS

The study has focused on the key parameters as environmental, social, and governance (ESG). The study has also compared the sustainability performance with Borsa Istanbul Sustainability Index (BIST SI) which was launched in the year 2014. The sustainability parameters as Thomson Reuters sustainability index, ESG matrices, Asset4 ESG, KLD's sustainability index, and Innovest. The study highlights that the better parameters are those that incorporate international institutions and are highly used globally these are Dow Jones, Standard & Poor's, and FTSE (The Financial Times Stock Exchange Group) launched in 2016.

Chandra and Kumar (2020) have conducted a study on the performance parameters and verified several possible indicators for sustainability measures and used a balanced scorecard.

KEY LEARNINGS

The study has used the balanced scorecard, the indicators of sustainability were measured in four different categories these are: internal process, learning, and growth, customer, and finance. Here, the sustainability performance was categorized into three sections economic, ecological, and Social. The assessment was based on the results of exploratory factor analysis and structural equation modeling. The significant indicators were used as training and education, management of diversity, health, and safety, equality among employees, dealing with discrimination issues with local communities, social employment, social assessment, emission measures, waste handling, social economic and environmental compliance, and assessment of indirect impacts on economic performance.

Giannakis et al. (2020) have developed a framework for sustainability performance measurement that works for the evaluation and selection of suppliers. The framework was developed using Analytic Network Process (ANP) method.

KEY LEARNINGS

The study has used the sustainability measures based on the global reporting initiative (GRI), the world business council for sustainable development (WBCSD), Dow Jones sustainability index, and the organization for economic cooperation and development (OECD). The significant parameters of sustainability found in the study were environmental parameters (greenhouse gas emission, consumption levels of energy and water, amount of waste generation), social (community care, community complaint handling, health and safety measures, employee working hours), economic (productivity, efficiency, return on investment, economic value and other investment on a sustainable production system).

Journeault et al. (2020) have studied sustainability performance reporting. In this research, the sustainability indicators were discussed.

KEY LEARNINGS

The study has highlighted the Global Reporting Initiatives (GRI). GRI guidelines have become the most significant reporting standards across the globe. It is one of the highly acceptable initiatives globally. The GRI contains three key parameters like economic, environmental, and social. But it is also having several critics associated with the standardization that creates challenges.

Lu et al. (2020) have conducted a study on critical factors that are influencing sustainable development performance. The study was about the electrical wire and cable industry.

KEY LEARNINGS

The study has mentioned the significance of the electrical wire and cable industry in terms of energy efficiency that leads to sustainability. The electrical wire and cable industry have potential contributions to sustainability in terms of energy efficiency. Multiple attribute decision-making (MADM) approaches including fuzzy DEMATEL, fuzzy DANP, and fuzzy modified VIKOR methods were used. The result of the study suggests that there is a need for improvement in environmental aspects for enhancing sustainability performance. Secondly, top management support is the key attribute in enhancing sustainability performance. The indicators were classified in the study as an economic aspect, environmental aspect, societal aspect, and organizational aspect.

Papoutsi and Sodhi (2020) have highlighted the sustainability performance and identified 51 sustainability indicators based on previous research. The study also focuses on reporting guidelines for sustainability.

KEY LEARNINGS

The study informs that several organizations disclose their sustainability-related activities with the help of sustainability reports. This information is useful for investors and stakeholders. The research has discussed sustainability reports and authenticity. The sustainable performance can be measured by Bloomberg's environmental and social governance (ESG) ratings and Dow Jones Sustainability Indices (DSJI). These methods were tested with the help of regression and logistic regression analysis. The guidelines as the Global Reporting Initiative (GRI) or the UN Global Compact deals with reporting on sustainability activities. There are a few third parties that rates sustainability performance. Firms like Morgan Stanley Capital Index (MSCI) Thomson Reuters Bloomberg helps in providing information on sustainability disclosure with the help of ESG.

9.3 METHODOLOGY

Since, the study has found key research gap that there are fewer studies that highlight the sustainability performance measurement methods, parameters, and challenges. Therefore, the study has adopted an exploratory research design (Mason et al., 2010) to enrich the theoretical knowledge in this research topic. The study is based on a critical review of the literature based on the objective of the study. After, the careful study of the previous literature the key learnings were narrated in the literature review section, and significant points based on the objective of the study were identified and tabulated in Table 9.1.

9.4 THEORETICAL FINDINGS AND DISCUSSION

In the study, 29 articles were studied based on the sustainability performance measurement methods, measurement parameters/indicators, and challenges associated with the measurement. All these articles were chosen in the study published between the years 1995-2020. The critical review of the article has given significant findings based on the sustainability measurement methods, measurement indicators, and associated challenges. The key learnings from the literature based on the research objective are presented in Table 9.1. The study has found several methods for measuring sustainability performance. Most of the organizations have a belief in Global Reporting Initiatives (GRI) for sustainability reporting which was proposed by the United Nations. There are several reporting patterns designed by authorities of the respective countries. But, some of them are only used in the home country rather than at the international level such as Borsa Istanbul Sustainability Index (BIST SI), etc. ESAs, ENVIO tables, ISEW, and Welsh Assembly Government Sustainability Indicators are some of the measurement methods studied in the literature. Several kinds of comparative methods among the indicators are also one of the significant methods found in the study. The major parameters or indicators for measuring sustainability are based on the triple bottom line approach i.e., people profit and planet that can be derived as a society, economy, and environment. All the indicators are based on the triple bottom line approach. Most of the indicators are set by institutions such as OECD, UNEP, World Bank, UNCSD, etc. The indicators are set and modified according to the nature of the industry. For example, the indicators used in the chemical industry will not work in the energy sector. There are several challenges associated with the performance measurement of the sustainability found in the study.

Findings of the study, indicate that there are challenges to measuring sustainability as it contains socio-economic issues, problem in acquiring data from sources, issues in fixing indicators, lack of comparative scales and standardization, etc. Moreover, in some cases measurement requires higher scientific knowledge. Both, qualitative methods and quantitative methods are available for sustainability measurement with their own pros and cons.

TABLE 9.1

Key Learnings from the Literature Based on the Research Objective

Author(s)	Sustainability Measurement Methods	Sustainability Measurement Parameters/Indicators	Sustainability Measurement Challenges
Hammond and World Resources Institute (1995)	Reporting based method using a systematic approach.	Indicators were suggested by OECD, & UNEP For example, climate change, urban environmental quality, etc. World Bank suggested: source indicators, support indicators, life support indicators, and human impact indicators	–
Gallopin (1996)	System approach.	Environmental standards as sustainability indicators. Key indicators: environment, finance, and society	–
Yeh and Li (1997)	Monitoring and evaluation with integrated remote sensing and geographical information systems (GIS) approach.	Environmental and social and economic parameters.	Socio-economic challenges hampering the sustainability goals of the region.
Azapagic and Perdan (2000)	Global Reporting Initiatives (GRI)	indicators are economic, social, environmental, financial, and other ethical parameters	–
Labuschagne et al. (2005)	Multi-criterion decision analysis (MCDA), valuation route" and "qualitative route".	GRI indicators, UNCSD Framework, IchemE, and WSI indicators	–
Hueting and Reijnders (2004)	SEEA and other sustainable development goals	Employees and their working environment, yearly withdrawal of water resources, carbon emission, etc.	–
Gallego (2006)	Methods based on GRI	Economic environmental, social, human rights, strategy management society, and product responsibility	–

(Continued)

TABLE 9.1 *(Continued)*
Key Learnings from the Literature Based on the Research Objective

Author(s)	Sustainability Measurement Methods	Sustainability Measurement Parameters/Indicators	Sustainability Measurement Challenges
Hezri and Dovers (2006)	Biophysical accounts, eco-efficiency, state of the environment reporting, and other methodological approaches guided and reported to the Government are used for sustainability performance.	Appropriate exploration of historical indicators and applications such as the study of public administration, environmental studies, and urban development studies	Acquiring legitimate data
Munday and Roberts (2006)	ESAs, ENVIO tables, ISEW, and Welsh Assembly Government Sustainability Indicators.	Indicators were major include housing, crime, education, air quality, employment, wild-life, waste, and ecological footprint, etc.	–
Bagheri and Hjorth (2007)	General Reporting methods based on the urban water system.	Indicators were based on the urban water system, sewage system stormwater, soil, air, etc.	–
Evans et al. (2009)	GRI for energy	energy conservation, greenhouse gas emission, renewable resources, price electricity, water, land and social impacts, etc.	–
Ramos and Caeiro (2010)	UN-based PSR frameworks.	Good practice factors, meta performance evaluation indicators	–
Hermans et al. (2011)	Contextual based factors are used for monitoring sustainability.	Socio-economic, ecological, structural characteristics, and Stakeholders based parameters.	
Shen et al. (2011)	Comparative analysis	Urban sustainability indicators & IUSIL	Issues related to fixing indicators by the institutions
Vithayasrichareon et al. (2012)	3As (accessibility, availability, and acceptability) which are used as a	electricity prices, electricity bills, subsidy reliability, supply time, reliability operating standards, etc.	Fixation of operating cost.

TABLE 9.1 *(Continued)*
Key Learnings from the Literature Based on the Research Objective

Author(s)	Sustainability Measurement Methods	Sustainability Measurement Parameters/Indicators	Sustainability Measurement Challenges
	sustainability analytical framework.		
Turcu (2013)	Methods based on GRI	Environmental, economic social, and institutional	–
Yin et al. (2014).	Eco-efficiency with positive matrix factorization, factor analysis, principal components analysis, and data envelopment analysis is also useful.	Based on financial and ecological information.	Accuracy in information gathering.
Feil et al. (2015)	General Reporting methods	social, economic, and environmental	–
Jurigová and Lencsésová (2015)	Reporting based on Document Agenda 21	Environmental indicators, economic indicators, and social indicators considering tourism industry	Societal challenges and changes
Rickels et al. (2016)	Reporting based on SDGs	17 (SDGs) based on environment, economy, and society.	–
Moldan et al. (2012)	*"Distance to target"* assessment model and *"decoupling"* by OECD	Waste disposal, air pollution, natural resources, climate change, etc.	Comparative issues.
Fazlagić and Skikiewicz (2019)	Sustainable reporting	Transportation connectivity, education level, universities, etc.	Challenges related to regional development.
Albawab et al. (2020)	MCDM-model and SWARA/ARAS	environmental, economical, resource-based, technological, and social, etc.	More scientific knowledge is required.
Aksoy et al (2020)	Comparative methods	ESG, BIST SI, Thomson Reuters sustainability index, Innovest Dow Jones, Standard & Poor's, and FTSE.	Lack of comparative scales
Chandra and Kumar (2020)	Balanced scorecard, Exploratory factor	Economic, ecological, and Social. Training and education, management	–

(Continued)

TABLE 9.1 *(Continued)*

Key Learnings from the Literature Based on the Research Objective

Author(s)	Sustainability Measurement Methods	Sustainability Measurement Parameters/Indicators	Sustainability Measurement Challenges
	analysis and structural equation modeling.	of diversity, health and safety, etc.	
Giannakis et al. (2020)	ANP, GRI), WBCSD, Dow Jones sustainability index	Environmental, social and economic	–
Journeault, Levant, and Picard (2020)	Sustainability performance reporting and GRI.	Economic, environmental and social	Standardization creates challenges.
Lu et al. (2020)	MADM approach, fuzzy DEMATEL, fuzzy DANP, and fuzzy modified VIKOR methods were used.	Economic aspect, environmental aspect, societal aspect and organizational aspect.	–
Papoutsi and Sodhi (2020)	Regression and logistic regression analysis and GRI	Bloomberg's environmental and social governance (ESG) ratings and Dow Jones Sustainability Indices (DSJI).	–

9.5 CONCLUSION

The number of articles shows that there is a significant gap in this research area. Though, the terminology sustainability and sustainable development are the top-notch keywords among the researchers. The present work has put an endeavour towards fulfilling the available research gap. The research is increasing towards how to establish sustainability in the industries by implementing several green initiatives in the business model. But, it is also important whether sustainability is taking place or not, and for that, there is a need of understanding the indicators and parameters for measuring the performance of sustainable development. The present chapter has highlighted several aspects related to the performance measurement of sustainability. The study suggests that there is a need for more attention in removing associated challenges towards measuring sustainability. There is a requirement for more international standards that industries can follow for measuring sustainability in firms. As of now, complexities are involved associated with measurement methods and indicators. It seems that industries are also confused about the use of sustainability performance measures and setting indicators. There is a need of more transparency towards these sustainability measures. The exclusivity of the present study is bringing all the measurement issues into a single study.

9.6 IMPLICATION OF THE STUDY

The outcome of the study can be beneficial for society in understanding the issues related to the measurement of sustainability performance. The managers can get the idea based on the research as the study has put all the measurement elements at one platform. The study can give a better understanding of sustainability measures for the decision-makers.

9.7 FUTURE RESEARCH SCOPE

This study has not covered all the things connecting to sustainability measures because of the time and resource-based limitations. But, this study can lead to another study in near future based on its findings. There is a need to study sustainability performance measurement with the help of real cases based on firms. This will enlighten about the ground reality of sustainability performance measurement.

REFERENCES

Aksoy, M., Yilmaz, M.K., Tatoglu, E., and Başar, M. (2020). Antecedents of corporate sustainability performance in Turkey: The effects of ownership structure and board attributes on non-financial companies. *Journal of Cleaner Production*, 276, Art. no. 124284.

Albawab, M., Ghenai, C., Bettayeb, M., and Janajreh, I. (2020). Sustainability performance index for ranking energy storage technologies using multi-criteria decision-making model and hybrid computational method. *Journal of Energy Storage*, 32, Art. no. 101820.

Azapagic, A., and Perdan, S. (2000). Indicators of sustainable development for industry: A general framework. *Process Safety and Environmental Protection*, 78(4), pp. 243–261.

Bagheri, A., and Hjorth, P. (2007). A framework for process indicators to monitor for sustainable development: Practice to an urban water system. *Environment, Development and Sustainability*, 9(2), pp. 143–161.

Bell, S., and Morse, S. (2012). *Sustainability indicators: Measuring the immeasurable?* NY, USA Routledge.

Chandra, D., and Kumar, D. (2020). Identifying key performance indicators of vaccine supply chain for sustainable development of mission indradhanush: A structural equation modeling approach. *Omega*, 101, Art. no. 102258.

Elkington, J. (1997). The triple bottom line. In *Environmental Management: Readings and Cases*, (Vol. 2). California, USA: Sage.

Evans, A., Strezov, V., and Evans, T.J. (2009). Assessment of sustainability indicators for renewable energy technologies. *Renewable and Sustainable Energy Reviews*, 13(5), pp. 1082–1088.

Fazlagić, J., and Skikiewicz, R. (2019). Measuring sustainable development-the creative economy perspective. *International Journal of Sustainable Development & World Ecology*, 26(7), pp. 635–645.

Feil, A.A., de Quevedo, D.M., and Schreiber, D. (2015). Selection and identification of the indicators for quickly measuring sustainability in micro and small furniture industries. *Sustainable Production and Consumption*, 3, pp. 34–44.

Gallopin, G.C. (1996). Environmental and sustainability indicators and the concept of situational indicators. A systems approach. *Environmental Modeling & Assessment*, 1(3), pp. 101–117.

Gallego, I. (2006). The use of economic, social and environmental indicators as a measure of sustainable development in Spain. *Corporate Social Responsibility and Environmental Management*, 13(2), pp. 78–97.

Giannakis, M., Dubey, R., Vlachos, I., and Ju, Y. (2020). Supplier sustainability performance evaluation using the analytic network process. *Journal of Cleaner Production*, 247, Art. no. 119439.

Hammond, A., and World Resources Institute (1995). *Environmental indicators: A systematic approach to measuring and reporting on environmental policy performance in the context of sustainable development* (Vol. 36). Washington, DC: World Resources Institute.

Hermans, F.L., Haarmann, W.M., & Dagevos, J.F. (2011). Evaluation of stakeholder participation in monitoring regional sustainable development. *Regional Environmental Change*, 11(4), pp. 805–815.

Hezri, A.A., and Dovers, S.R. (2006). Sustainability indicators, policy and governance: Issues for ecological economics. *Ecological Economics*, 60(1), pp. 86–99.

Huang, H.F., Kuo, J., and Lo, S.L. (2011). Review of PSR framework and development of a DPSIR model to assess greenhouse effect in Taiwan. *Environmental Monitoring and Assessment*, 177(1–4), pp. 623–635.

Hueting, R., and Reijnders, L. (2004). Broad sustainability contra sustainability: the proper construction of sustainability indicators. *Ecological Economics*, 50(3-4), pp. 249–260.

Journeault, M., Levant, Y., and Picard, C.F. (2020). Sustainability performance reporting: A technocratic shadowing and silencing. *Critical Perspectives on Accounting*, 74, Art. no. 102145.

Jurigová, Z., and Lencsésová, Z. (2015). Monitoring system of sustainable development in cultural and mountain tourism destinations. *Journal of Competitiveness*, 7(1), pp. 35–52.

Kushwaha, G.S., and Sharma, N.K. (2016). Green initiatives: A step towards sustainable development and firm's performance in the automobile industry. *Journal of Cleaner Production*, 121, pp. 116–129.

Labuschagne, C., Brent, A.C., and Van Erck, R.P. (2005). Assessing the sustainability performances of industries. *Journal of Cleaner Production*, 13(4), pp. 373–385.

Laosirihongthong, T., Samaranayake, P., Nagalingam, S.V., and Adebanjo, D. (2020). Prioritization of sustainable supply chain practices with triple bottom line and organizational theories: Industry and academic perspectives. *Production Planning & Control*, 31(14), pp. 1207–1221.

Lélé, S. M. (1991). Sustainable development: A critical review. *World Development*, 19(6), pp. 607–621.

Lekan, A., Aigbavboa, C., Babatunde, O., Olabosipo, F., and Christiana, A. (2020). Disruptive technological innovations in construction field and fourth industrial revolution intervention in the achievement of the sustainable development goal 9. *International Journal of Construction Management*, pp. 1–12. https://doi.org/10.1080/15623599.202 0.1819522

Lu, M.T., Tsai, J.F., Shen, S.P., Lin, M.H., and Hu, Y.C. (2020). Estimating sustainable development performance in the electrical wire and cable industry: Applying the integrated fuzzy MADM approach. *Journal of Cleaner Production*, 277, Art. no. 122440.

Lynch, M.J., Barrett, K.L., Stretesky, P.B., and Long, M.A. (2016). The weak probability of punishment for environmental offenses and deterrence of environmental offenders: A discussion based on USEPA criminal cases, 1983–2013. *Deviant Behavior*, 37(10), pp. 1095–1109.

Mason, P., Augustyn, M., and Seakhoa-King, A. (2010). Exploratory study in tourism: Designing an initial, qualitative phase of sequenced, mixed methods research. *International Journal of Tourism Research*, 12(5), pp. 432–448.

Moldan, B., Janoušková, S., and Hák, T. (2012). How to understand and measure environmental sustainability: Indicators and targets. *Ecological Indicators*, 17, pp. 4–13.

Munday, M., and Roberts, A. (2006). Developing approaches to measuring and monitoring sustainable development in Wales: A review. *Regional Studies*, 40(5), pp. 535–554.

Ness, B., Urbel-Piirsalu, E., Anderberg, S., and Olsson, L. (2007). Categorising tools for sustainability assessment. *Ecological Economics*, 60, pp. 498–508.

Papoutsi, A., and Sodhi, M.S. (2020). Does disclosure in sustainability reports indicate actual sustainability performance? *Journal of Cleaner Production*, 260, Art. no. 121049.

Ramos, T.B., and Caeiro, S. (2010). Meta-performance evaluation of sustainability indicators. *Ecological Indicators*, 10(2), pp. 157–166.

Reed, M.S., Fraser, E.D.G., and Dougill, A.J. (2006). An adaptive learning process for developing and applying sustainability indicators with local communities. *Ecological Economics*, 59(4), pp. 406–418.

Rickels, W., Dovern, J., Hoffmann, J., Quaas, M.F., Schmidt, J.O., and Visbeck, M. (2016). Indicators for monitoring sustainable development goals: An application to oceanic development in the European Union. *Earth's Future*, 4(5), pp. 252–267.

Sharma, N.K., and Kushwaha, G.S. (2018). Factors affecting towards green purchase behavior among young consumers in India. In *Green Computing Strategies for Competitive Advantage and Business Sustainability* (pp. 183–210). Pennsylvania, USA: IGI Global.

Shen, L.Y., Ochoa, J.J., Shah, M.N., and Zhang, X. (2011). The application of urban sustainability indicators–A comparison between various practices. *Habitat International*, 35(1), pp. 17–29.

Singh, R.K., Murty, H.R., Gupta, S.K., and Dikshit, A.K. (2012). An overview of sustainability assessment methodologies. *Ecological Indicators*, 15(1), pp. 281–299.

Turcu, C. (2013). Re-thinking sustainability indicators: Local perspectives of urban sustainability. *Journal of Environmental Planning and Management*, 56(5), pp. 695–719.

Varma, V.S., and Kalamdhad, A.S. (2017). Solid waste: Environmental threats and management. In *Environmental Pollutants and Their Bioremediation Approaches*. FL, USA: CRC Press, pp. 337–368.

Vithayasrichareon, P., MacGill, I.F., and Nakawiro, T. (2012). Assessing the sustainability challenges for electricity industries in ASEAN newly industrialising countries. *Renewable and Sustainable Energy Reviews*, 16(4), pp. 2217–2233.

WCED (1987). *Our common future. Brundtland report. World Commission on Environment and Development* (pp. 1–300). Brundtland: Oxford University Press.

Yeh, A.G.O., and Li, X. (1997). An integrated remote sensing and GIS approach in the monitoring and evaluation of rapid urban growth for sustainable development in the Pearl River Delta, China. *International Planning Studies*, 2(2), pp. 193–210.

Yin, K., Wang, R., An, Q., Yao, L., and Liang, J. (2014). Using eco-efficiency as an indicator for sustainable urban development: A case study of Chinese provincial capital cities. *Ecological Indicators*, 36, pp. 665–671.

Zoeteman, K. (2004). Sustainability of nations. In Boersema, J.J., and Reijnders, L. (Eds.), *Principles of environmental science*. Dordrecht: Kluwer Academic Publishers.

10 A Systematic Review of Industrial Sustainability Research

Gaurav Kumar Badhotiya
Department of Mechanical Engineering, Graphic Era
(Deemed to be University), Dehradun, Uttarakhand, India

10.1 INTRODUCTION

The prevention of pollution, reduction and management of waste, and sustainable manufacturing policies are the focus points of modern industry. Most industrial activities are deteriorating the environment by excessive consumption of natural resources (Cioca et al. 2019). Sustainable development is the key to ensure the prevention and control of risk and pollution in industrial organizations. Sustainability and sustainable development both follow the Triple bottom line approach concerning social, environmental and economic aspects (Pathak et al. 2020). According to World Commission on Environment and Development (Brundtland, 1987), "sustainable development is the development that meets the needs of the present without compromising the ability of the future generations to meet their own needs".

Sustainable development led the foundation of industrial sustainability (Fang et al. 2007). Industrial sustainability is related to adoption of business strategies focusing on human and natural environment concerning present as well as future generations. Several authors have defined industrial sustainability, the details of which are given in Table 10.1.

There exist few articles that have reviewed the industrial sustainability in different context. Bull (2001) has reviewed the literature of biotechnology for industrial sustainability. Several case studies have been presented to demonstrate application of biotechnology as a clean technology. Arena et al. (2009) discussed operational view of IS based on definitions, tools and indicators. Further in 2013, Tonelli et al. defined and provided direction for transformation towards IS. Performance measurement and organization issues were also discussed along with recommendation to educators and policy makers. Smart et al. (2017) provided a systematic review of IS between 1992 to 2014. Descriptive and in depth thematic analysis was performed to identify prominent dialogues in the literature. Moreover, Feil et al. (2019) reviewed literature to identify sustainability indicators, benefits and limitations of IS. The study considered 24 articles from 1998 to 2018 for

DOI: 10.1201/9781003102304-10

TABLE 10.1
Definitions of Industrial Sustainability

Definition	Citation
"Intended to mean both the interaction of global industrial civilization with the natural environment and the aggregate of opportunities for individual industries to transform their relationships with the natural environment".	Socolow et al. (1997)
"Conceptualization, design and manufacture of goods and services that meet the needs of the present generation while not diminishing economic, social and environmental opportunity in the long term".	Jansson et al. (2000)
"Industrial sustainability can be defined as the adoption of business strategies and activities that answer the actual demands of enterprises and shareholders, by protecting, sustaining and improving the human and natural resources that will be necessary for future improvements".	Labuschagne and Brent (2007)
"Industrial sustainability focuses on how to pursue the short- to long-term sustainable development of an industrial system, such as a plant, corporation, geographic region, industrial zone, or beyond, where material and energy efficiencies, waste reduction, safety, and synergies among the systems, etc., are among the major concerns".	Piluso et al. (2010)
"Industrial sustainability refers to the end state of a transformation process where industry is part of, and actively contributing to, a socially, environmentally and economically sustainable planet".	Tonelli et al. (2013)

analysis and proposed a framework of IS. Javanmardi et al. (2020) reviewed article that have applied Grey system theory approach in the domain of sustainability. IS was the one among the ten different areas in which the approach was applied. Although existing literature review have provided much information regarding concept and implementation of IS, more information is required to understand the level of analysis, methodologies adopted, analysis techniques used and application of IS.

The objective of current study is to consolidate the literature, illuminate the status of research and future avenues on industrial sustainability. Next section explains the review methodology adopted in this study for structured literature review. Section 10.3 presents the results and discussion to highlight research gaps. The last section concludes this study with few future research suggestions for the academicians and industry personals working in this area.

10.2   REVIEW METHODOLOGY

A literature review is an essential activity of research with an aim to find research gaps in the area considered. The Structured literature review is a verifiable and rigorous methodology which reduces the biasness of the results. The review is performed by searching the relevant literature using suitable keywords and conducting the analysis to draw out meaningful results. In this study, literature is collected from Scopus database which contains publications from several reputed scientific databases such as Elsevier, Emerald Insight, Taylor and Francis, Springer, IEEE and MDPI.

Initial search in "Abstract, title and keywords" gives 219 results. The publications were selected by excluding books, thesis, editorial notes, and white papers. The remaining articles were read in full to identify the relevance to the topic. After looking at relevance and cross referencing, 58 articles from 2001 to 2020(Q3) were finalized. The selected articles were then classified and analysed based on growth of the field, publisher, journal, methodology adopted, analysis technique, level, application and geographical focus of the study. On the basis of results, research gaps, findings and future research avenues are discussed.

10.3   RESULTS AND DISCUSSION

Figure 10.1 demonstrates the number of papers published on IS year wise. It can be observed from the figure that the number of articles is increasing yearly which shows the continuous growth and increasing attention from researchers on this field.

The frequency of publication according to the publisher is shown by Figure 10.2. The highest number of articles was published in Elsevier, followed by Springer, MDPI, and Wiley. The results indicate that the field is covered by different publishers.

The frequency of publication according to journal, book, and conference proceedings is shown in Table 10.2. *Journal of Cleaner Production*, which is a highly reputed journal in the area of environmental research, is leading in this field. The articles published in particular in *Journal of Cleaner Production, Sustainability, Procedia CIRP, Procedia Manufacturing, Progress in Industrial Ecology – An International Journal, Studies in Computational Intelligence,* covers 51.7% of the

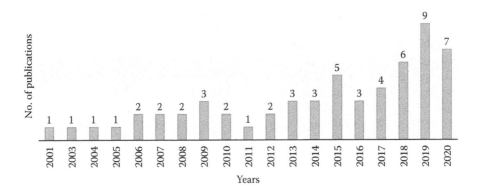

FIGURE 10.1   Frequency of publication by year.

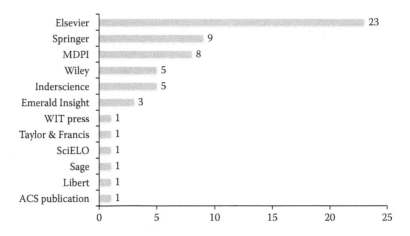

FIGURE 10.2   Frequency of publications by publisher.

publication identified. These journals are either among the most qualified journals or the reputed conference proceedings.

The implementation of IS can be categorized at three levels viz. firm, country and global. Firm level indicates the activities performed or suggested at one industry or a type of industry segment. The country level indicates the IS practice and policy implementation focusing a country. The third level is related to general understanding, concept and promotion of IS activities and policies. It can be seen from Figure 10.3 that country level analysis (48%) is having the maximum percentage followed by global (37%) and firm level (15%).

Figures 10.4 and 10.5 shows the distribution of methodology and analysis techniques adopted by authors working on IS. The conceptual paper discuss about ideas, theories, understanding and promotion of to achieve IS. Empirical studies present surveys and case studies that are focused on measurable IS practices and indicators. The dominant research methodology implemented in the articles is Empirical (64%), followed by Conceptual (26%) and Review (10%).

In analysis technique, qualitative (64%) techniques is the most preferred one followed by quantitative (25%) and few other articles used mixed approach.

Table 10.3 classify articles based on their research focus, industry application and geographical focus. The categories of research focus are understanding, assessment, performance measurement, and implementation. The majority of articles are related to implementation (40.3%) of IS followed by understanding (25%) the concept. The industry level classification shows that the manufacturing sector (58.8%) is the most preferred. There exist several areas which are not yet addressed or very few articles have worked upon such as food, energy, and medical industry. In the classification based on geographical focus, majority of articles have focused on developing countries e.g., China is having maximum number of studies. The large number of studies on country level analysis shows the worldwide growth of the IS field.

TABLE 10.2
Frequency of Publication According to Journal/Book/Proceeding

S.no.	Journal/Book/Conference Proceeding	Frequency
1	Journal of Cleaner Production	10
2	Sustainability	8
3	Procedia CIRP	5
4	Procedia Manufacturing	3
5	Progress in Industrial Ecology – An International Journal	2
6	Studies in Computational Intelligence	2
7	AiChE	1
8	Applied Biochemistry and Biotechnology	1
9	Clean Technology and Environmental Policy	1
10	Energy Policy	1
11	Eng. Life Sci	1
12	Foresight	1
13	Gestão & Produção	1
14	Industrial Engineering and Chemistry Research	1
15	International Environmental Agreements	1
16	International Journal of Energy Sector Management	1
17	International Journal of Environmental Technology and Management	1
18	International Journal of Sustainable Engineering	1
19	Journal of Environmental Management	1
20	Journal of Industrial Ecology	1
21	Journal of the Knowledge Economy	1
22	Land Use Policy	1
23	Measurement and Control	1
24	R&D Management	1
25	Re-engineering Manufacturing for Sustainability	1
26	Service Orientation in Holonic and Multi-agent Manufacturing	1
27	Sustainability: The Journal of Record	1
28	Sustainable Development	1
29	Sustainable Production and Consumption	1
30	Transactions on Ecology and the Environment	1
31	Korean Journal of Chemical Engineering	1
32	International Journal of Product Lifecycle Management	1
33	International Journal of Business Innovation and Research	1
34	International Journal of Operations & Production Management	1

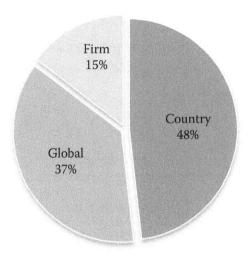

FIGURE 10.3   Level of analysis.

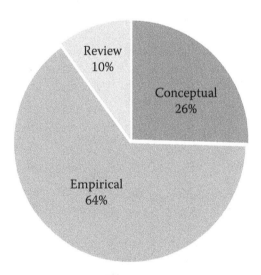

FIGURE 10.4   Research methodology adopted by authors.

10.4   CONCLUSION AND FUTURE RESEARCH DIRECTIONS

The objective of this study is to provide a structured literature review and analysis of 58 articles published in the field of industrial sustainability between the period of 2001 to 2020(Q3). During the period, industrial sustainability field have emerged from one article in 2001 to nine articles in 2019 and seven articles in 2020(Q3). The articles are categorized on several criteria to interpret meaningful results and research gaps. The outcome of this study suggests to focus more on firm level analysis and application of quantitate techniques. Moreover, research

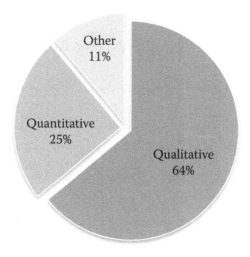

FIGURE 10.5   Analysis techniques adopted by authors.

TABLE 10.3
Article Classification Based on Research Focus, Industry, and Geographical Focus

S.no.	Author	Year	Research Focus	Industry	Geographical Focus
1	Ruiz-Puente and Jato-Espino	2020	Implementation	Industrial park	Spain
2	Quartey and Oguntoye	2020	Understanding	NA	Africa
3	Liu et al.	2020	Implementation	Agriculture	China
4	Moskovich	2020	Understanding	Agriculture	Israel
5	Shah et al.	2020	Assessment	NA	Ulsan, Korea
6	Tolstykh et al.	2020	Assessment	NA	Kalundborg, Baltic
7	Cioca et al.	2019	Implementation	Automotive	Romania
8	Arbolino and Simone	2019	Performance Measurement	NA	Europe
9	Trianni et al.	2019	Performance Measurement	Manufacturing	Italian and German
10	Riesener et al.	2019	Implementation	Manufacturing	NA
11	Gudukeya et al.	2019	Implementation	Manufacturing	Zimbabwe
12	Cagno et al.	2019	Performance Measurement	Manufacturing	Italy
13	Demartini et al.	2019a	Implementation	Manufacturing	NA
14	Chowdhury et al.	2019	Implementation	NA	Bangladesh

(Continued)

TABLE 10.3 (Continued)
Article Classification Based on Research Focus, Industry, and Geographical Focus

S.no.	Author	Year	Research Focus	Industry	Geographical Focus
15	Demartini et al.	2019b	Implementation	NA	China
16	Amaral et al.	2018	Implementation	Textile	Brazil
17	Arbolino et al.	2018	Assessment	NA	Italy
18	Neri et al.	2018	Implement`ation	Manufacturing	Italy
19	Wang et al.	2018	Assessment	NA	NA
20	Demartini et al.	2018	Performance Measurement	Food	NA
21	Trianni et al.	2017	Implementation	Manufacturing	Italy
22	Oguntoye and Evans	2017	Understanding	Manufacturing	Africa (Nigeria)
23	Shankar et al.	2017	Understanding	Manufacturing	India
24	Davé et al.	2016	Performance Measurement	Furniture industry	NA
25	Ibiyemi et al.	2016	Assessment	NA	NA
26	Tonelli et al.	2016	Understanding	Manufacturing	NA
27	Tan et al.	2015	Assessment	Manufacturing	Singapore
28	Feil et al.	2015	Performance Measurement	Furniture industry	Brazil
29	L¢pez-Morales et al.	2015	Understanding	NA	NA
30	Hauschild	2015	Assessment	NA	NA
31	Rocco	2015	Understanding	NA	NA
32	Olinto	2014	Assessment	Manufacturing	NA
33	Cagno et al.	2014	Implementation	Manufacturing	Netherlands
34	Lunt et al.	2014	Implementation	Manufacturing	UK
35	Despeisse et al.	2013	Understanding	NA	NA
36	Chan and Santos-Paulino	2013	Implementation	Energy	China
37	Liu and Huang	2012	Implementation	Manufacturing	NA
38	Despeisse et al.	2012	Understanding	Manufacturing	NA
39	Overcash and Twomey	2011	Understanding	NA	United States
40	Piluso et al.	2010	Assessment	Manufacturing	NA
41	Ackom et al.	2010	Understanding	Energy	Canada
42	Erol et al.	2009	Assessment	Retail	Turky
43	Jacobsen	2006	Implementation	Manufacturing	Denmark, China
44	Jacobsen	2006	Implementation	NA	Denmark, China
45	Zeng et al.	2008	Assessment	Manufacturing	Shanghai, China
46	Edet et al.	2007	Understanding	NA	NA

TABLE 10.3 (Continued)
Article Classification Based on Research Focus, Industry, and Geographical Focus

S.no.	Author	Year	Research Focus	Industry	Geographical Focus
47	Fang et al.	2007	Implementation	Process (sugar, chemical)	China
48	Hermansen et al.	2006	Understanding	NA	NA
49	Ohtake et al.	2006	Implementation	Biotechnology	Japan
50	Baldwin et al.	2005	Implementation	Manufacturing	United Kingdom
51	Paramanathan et al.	2004	Implementation	NA	NA
52	Ortiz-Gallarza et al.	2003	Assessment	Refinery	Hidalgo, Maxico

should also focus on performance measurement and application apart from manufacturing area. Further analysis could be directed towards analysing the effect of current pandemic situation on policies of industrial sustainability. Identification of barriers and enablers for adoption of industrial sustainability measures is also a promising area of research.

REFERENCES

Ackom, E. K., Mabee, W. E., and Saddler, J. N. (2010). Industrial sustainability of competing wood energy options in Canada. *Applied Biochemistry and Biotechnology*, 162(8), pp. 2259–2272. https://doi.org/10.1007/s12010-010-9000-6

Arbolino, R., Boffardi, R., Lanuzza, F., and Ioppolo, G. (2018). Monitoring and evaluation of regional industrial sustainability: Evidence from Italian regions. *Land Use Policy*, 75(April), pp. 420–428. https://doi.org/10.1016/j.landusepol.2018.04.007

Arena, M., Ciceri, N. D., Terzi, S., Bengo, I., Azzone, G., and Garetti, M. (2009). A state-of-the-art of industrial sustainability: definitions, tools and metrics. *International Journal of Product Lifecycle Management*, 4(1–3), 207–251. https://doi.org/10.1504/IJPLM.2009.031674

Baldwin, J. S., Allen, P. M., Winder, B., and Ridgway, K. (2005). Modelling manufacturing evolution: Thoughts on sustainable industrial development. *Journal of Cleaner Production*, 13(9), pp. 887–902. https://doi.org/10.1016/j.jclepro.2004.04.009

Brundtland, G.H. (1987). What is sustainable development. Our common future, 8(9).

Bull, A.T., (2001). Biotechnology for industrial sustainability. *Korean Journal of Chemical Engineering*, 18(2), Art. no. 137. https://doi.org/10.1007/BF02698451

Cagno, E., Neri, A., Howard, M., Brenna, G., and Trianni, A. (2019). Industrial sustainability performance measurement systems: A novel framework. *Journal of Cleaner Production*, 230, pp. 1354–1375. https://doi.org/10.1016/j.jclepro.2019.05.021

Cagno, E., Trianni, A., Worrell, E., and Miggiano, F. (2014). Barriers and drivers for energy efficiency: Different perspectives from an exploratory study in the Netherlands. *Energy Procedia*, 61, pp. 1256–1260. https://doi.org/10.1016/j.egypro.2014.11.1073

Chen, S., and Santos-Paulino, A. U. (2013). Energy consumption restricted productivity re-estimates and industrial sustainability analysis in post-reform China. *Energy Policy*, 57, pp. 52–60. https://doi.org/10.1016/j.enpol.2012.08.060

Chowdhury, H., Chowdhury, T., Thirugnanasambandam M., Farhan, M., Ahamed, J. U., Saidur, R., and Sait, S. M. (2019). A study on exergetic efficiency vis-à-vis sustain-ability of industrial sector in Bangladesh. *Journal of Cleaner Production*, 231(2019), pp. 297–306. https://doi.org/10.1016/j.jclepro.2019.05.174

Cioca, L. I., Ivascu, L., Turi, A., Artene, A., and Gaman, G. A. (2019). Sustainable devel-opment model for the automotive industry. *Sustainability (Switzerland)*, 11(22), pp. 1–22. https://doi.org/10.3390/su11226447

Davé, A., Salonitis, K., Ball, P., Adams, M., and Morgan, D. (2016). Factory eco-efficiency modelling: Framework application and analysis. *Procedia CIRP*, 40, pp. 214–219. https://doi.org/10.1016/j.procir.2016.01.105

Demartini, M., Bertani, F., and Tonelli, F. (2019a). AB-SD hybrid modelling approach: A framework for evaluating industrial sustainability scenarios. In *Studies in Computational Intelligence* (vol. 803). Springer International Publishing. https://doi.org/10.1007/978-3-030-03003-2_17

Demartini, M., Evans, S., and Tonelli, F. (2019b). Digitalization technologies for industrial sustainability. *Procedia Manufacturing*, 33, pp. 264–271. https://doi.org/10.1016/j.promfg.2019.04.032

Demartini, M., Pinna, C., Aliakbarian, B., Tonelli, F., and Terzi, S. (2018). Soft drink supply chain sustainability: A case based approach to identify and explain best practices and key performance indicators. *Sustainability (Switzerland)*, 10(10), pp. 1–24. https://doi.org/10.3390/su10103540

Despeisse, M., Ball, P. D., Evans, S., and Levers, A. (2012). Industrial ecology at factory level - A conceptual model. *Journal of Cleaner Production*, 31, pp. 30–39. https://doi.org/10.1016/j.jclepro.2012.02.027

Despeisse, M., Ball, P. D., and Evans, S. (2013). Strategies and ecosystem view for industrial sustainability. *Re-Engineering Manufacturing for Sustainability - Proceedings of the 20th CIRP International Conference on Life Cycle Engineering*, Figure 2, pp. 565–570. https://doi.org/10.1007/978-981-4451-48-2_92

Do Amaral, M. C., Zonatti, W. F., Da Silva, K. L., Junior, D. K., Neto, J. A., and Baruque-Ramos, J. (2018). Industrial textile recycling and reuse in Brazil: Case study and considerations concerning the circular economy. *Gestao e Producao*, 25(3), pp. 431–443. https://doi.org/10.1590/0104-530X3305

Edet, E., Wright, D., and Street, G. (2007). The importance of the ecological metaphor in understanding industrial sustainability. *Measurement and Control*, 40(8), pp. 235–238. https://doi.org/10.1177/002029400704000801

Erol, I., Cakar, N., Erel, D., and Sari, R. (2009). Sustainability in the Turkish retailing in-dustry. *Sustainable Development*, 17(1), pp. 49–67. https://doi.org/10.1002/sd.369

Fang, Y., Côté, R. P., and Qin, R. (2007). Industrial sustainability in China: Practice and prospects for eco-industrial development. *Journal of Environmental Management*, 83(3), pp. 315–328. https://doi.org/10.1016/j.jenvman.2006.03.007

Feil, A. A., de Quevedo, D. M., and Schreiber, D. (2015). Selection and identification of the indicators for quickly measuring sustainability in micro and small furniture industries. *Sustainable Production and Consumption*, 3(June), pp. 34–44. https://doi.org/10.1016/j.spc.2015.08.006

Feil, A.A., Schreiber, D., Haetinger, C., Strasburg, V. J. and Barkert, C. L. (2019). Sustainability indicators for industrial organizations: Systematic review of literature. *Sustainability*, 11(3), Art. no. 854. https://doi.org/10.3390/su11030854

Gudukeya, L., Mbohwa, C., and Mativenga, P. T. (2019). Industrial sustainability in a challenged economy: The Zimbabwe steel industry. *Procedia Manufacturing*, 33, pp. 562–569. https://doi.org/10.1016/j.promfg.2019.04.070

Hauschild, M. Z. (2015). Better - but is it good enough? On the need to consider both eco-efficiency and eco-effectiveness to gauge industrial sustainability. *Procedia CIRP*, 29, pp. 1–7. https://doi.org/10.1016/j.procir.2015.02.126

Hermansen, J. E. (2006). Industrial ecology as mediator and negotiator between ecology and industrial sustainability. *Progress in Industrial Ecology*, 3(1–2), pp. 75–94. https://doi.org/10.1504/pie.2006.010042

Ibiyemi, A. O., Adnan, Y. M., and Daud, M. N. (2016). The validity of the classical Delphi applications for assessing the industrial sustainability-correction factor: An example study. *Foresight*, 18(6), pp. 603–624. https://doi.org/10.1108/FS-04-2016-0016

Jacobsen, N. B. (2006). Industrial symbiosis in Kalundborg, Denmark: A quantitative assessment of economic and environmental aspects. *Journal of Industrial Ecology*, 10(1–2), pp. 239–255. https://doi.org/10.1162/108819806775545411

Jacobsen, N. B. (2009). Exploring mechanisms for mobilising industrial sustainability models across different industrial locations. *International Journal of Environmental Technology and Management*, 10(3–4), pp. 442–457. https://doi.org/10.1504/IJETM.2009.023745

Jansson, P. M., Gregory, M. J., Barlow, C., Phaal, R., Farrukh, C. J. P., Probert, D. R., and So, V. (2000). *Industrial Sustainability – A Review of UK and International Research and Capabilities*. Cambridge: University of Cambridge

Javanmardi, E., Liu, S. and Xie, N., (2020). Exploring grey systems theory-based methods and applications in sustainability studies: A systematic review approach. *Sustainability*, 12(11), Art. no. 4437. https://doi.org/10.3390/su12114437

Labuschagne, C., and Brent, A.C., (2007). Sustainability assessment criteria for projects and technologies: Judgements of industry managers. *South African Journal of Industrial Engineering*, 18(1), pp. 19–33. https://hdl.handle.net/10520/EJC46143

Liu, Z., and Huang, Y. (2012). Technology evaluation and decision making for sustainability enhancement of industrial systems under uncertainty. *AIChE Journal*, 58(6), pp. 1841–1852. https://doi.org/10.1002/aic.13818

Liu, Z., Liang, H., Pu, D., Xie, F., Zhang, E., and Zhou, Q. (2020). How does the control of grain purchase price affect the sustainability of the national grain industry? One empirical study from China. *Sustainability (Switzerland)*, 12(5), pp. 1–21. https://doi.org/1 0.3390/su12052102

Lépez-Morales, V., Ouzrout, Y., Manakitsirisuthi, T., and Bouras, A. (2015). MKMSIS: A multi-agent knowledge management system for industrial sustainability. In *Artificial Intelligence Applications in Information and Communication Technologies*. Cham: Springer, pp. 195–213. https://doi.org/10.1007/978-3-319-19833-0_9

Lunt, P., Ball, P., and Levers, A. (2014). Barriers to industrial energy efficiency. *International Journal of Energy Sector Management*, 8(3), pp. 380–394. https://doi.org/10.1108/IJESM-05-2013-0008

Moskovich, Y. (2020). Family home business in Kibbutz industry sustainability. *Sustainability (Switzerland)*, 12(13), pp. 1–20. https://doi.org/10.3390/su12135388

Neri, A., Cagno, E., Di Sebastiano, G., and Trianni, A. (2018). Industrial sustainability: Modelling drivers and mechanisms with barriers. *Journal of Cleaner Production*, 194, pp. 452–472. https://doi.org/10.1016/j.jclepro.2018.05.140

Oguntoye, O., and Evans, S. (2017). Framing manufacturing development in Africa and the influence of industrial sustainability. *Procedia Manufacturing*, 8, pp. 75–80. https://doi.org/10.1016/j.promfg.2017.02.009

Ohtake, H., Yamashita, S., and Kato, J. (2006). Development of a new biotechnological basis for improving industrial sustainability in Japan. *Engineering in Life Sciences*, 6(3), pp. 278–284. https://doi.org/10.1002/elsc.200620124

Olinto, A. C. (2014). Vector space theory of sustainability assessment of industrial processes. *Clean Technologies and Environmental Policy*, 16(8), pp. 1815–1820. https://doi.org/1 0.1007/s10098-014-0729-4

Ortiz-Gallarza, S. M., Nava, M., Rodríguez, L., Barrera, A., and Villaseñor, E. (2003). Selection of environmental parameters to estimate an industrial sustainability index. *Sustainable World*, 6, pp. 363–372.

Overcash, M., and Twomey, J. M. (2011). Structure of industrial or corporate sustainability programmes. *International Journal of Sustainable Engineering*, 4(2), pp. 109–114. https://doi.org/10.1080/19397038.2011.558932

Paramanathan, S., Farrukh, C., Phaal, R., and Probert, D. (2004). Implementing industrial sustainability: The research issues in technology management. *R&D Management*, 34(5), pp. 527–537. https://doi.org/10.1111/j.1467-9310.2004.00360.x

Pathak, P., Singh, M. P., and Badhotiya, G. K. (2020). Performance obstacles in sustainable manufacturing – model building and validation. *Journal of Advances in Management Research*, 17(4), pp. 549–566. https://doi.org/10.1108/JAMR-03-2020-0031

Piluso, C., Huang, J., Liu, Z., and Huang, Y. (2010). Sustainability assessment of industrial systems under uncertainty: A fuzzy logic based approach to short-term to midterm predictions. *Industrial and Engineering Chemistry Research*, 49(18), pp. 8633–8643. https://doi.org/10.1021/ie100164r

Quartey, S. H., and Oguntoye, O. (2020). Understanding and promoting industrial sustainability in Africa through the triple helix approach: A conceptual model and research propositions. *Journal of the Knowledge Economy*, 1–19. https://doi.org/10.1007/s13132-020-00660-2

Riesener, M., Dölle, C., and Kuhn, M. (2019). Innovation ecosystems for industrial sustainability. *Procedia CIRP*, 80, pp. 27–32. https://doi.org/10.1016/j.procir.2019.01.035

Rocco, C. (2015). Aspects to enhance environmental sustainability in industry: A brief roadmap. *Sustainability (United States)*, 8(5), pp. 254–260. https://doi.org/10.1089/SUS.2015.29022

Ruiz-Puente, C., and Jato-Espino, D. (2020). Systemic analysis of the contributions of co-located industrial symbiosis to achieve sustainable development in an industrial park in Northern Spain. Sustainability, 12(14), 5802. https:// doi.org/10.3390/su12145802

Shah, I. H., Dong, L., and Park, H. S. (2020). Tracking urban sustainability transition: An eco-efficiency analysis on eco-industrial development in Ulsan, Korea. *Journal of Cleaner Production*, 262, Art. no.121286. https://doi.org/10.1016/j.jclepro.2020.121286

Shankar, Y. S., Mohan, D., Vaid, P., Anand, M., and Ganesh, S. (2017). Emerging approaches for industrial sustainability and feasible applications in India. *Progress in Industrial Ecology*, 11(1), pp. 61–78. https://doi.org/10.1504/PIE.2017.086141

Smart, P., Hemel, S., Lettice, F., Adams, R., and Evans, S., (2017). Pre-paradigmatic status of industrial sustainability: A systematic review. *International Journal of Operations & Production Management*, 37(10), pp. 1425–1450. https://doi.org/10.1108/IJOPM-02-201 6-0058

Socolow, R., Andrews, C., Berkhout, F., and Thomas, V., eds. (1997). *Industrial Ecology and Global Change* (vol. 5). United Kingdom: Cambridge University Press.

Tan, H. X., Yeo, Z., Ng, R., Tjandra, T. B., and Song, B. (2015). A sustainability indicator framework for Singapore small and medium-sized manufacturing enterprises. *Procedia CIRP*, 29, pp. 132–137. https://doi.org/10.1016/j.procir.2015.01.028

Tolstykh, T., Shmeleva, N., and Gamidullaeva, L. (2020). Evaluation of circular and integration potentials of innovation ecosystems for industrial sustainability. *Sustainability (Switzerland)*, 12(11), pp. 1–17. https://doi.org/10.3390/su12114574

Tonelli, F., Evans, S., and Taticchi, P., (2013). Industrial sustainability: Challenges, perspectives, actions. *International Journal of Business Innovation and Research*, 7(2), pp. 143–163. https://doi.org/10.1504/IJBIR.2013.052576

Tonelli, F., Fadiran, G., Raberto, M., and Cincotti, S. (2016). Approaching industrial sustainability investments in resource efficiency through agent-based simulation. *Studies in Computational Intelligence*, 640, pp. 145–155. https://doi.org/10.1007/978-3-319-30337-6_14

Trianni, A., Cagno, E., and Neri, A. (2017). Modelling barriers to the adoption of industrial sustainability measures. *Journal of Cleaner Production*, 168, pp. 1482–1504. https://doi.org/10.1016/j.jclepro.2017.07.244

Trianni, A., Cagno, E., Neri, A., and Howard, M. (2019). Measuring industrial sustainability performance: Empirical evidence from Italian and German manufacturing small and medium enterprises. *Journal of Cleaner Production*, 229, pp. 1355–1376. https://doi.org/10.1016/j.jclepro.2019.05.076

Wang, C., Wang, L., and Dai, S. (2018). An indicator approach to industrial sustainability assessment: The case of China's Capital Economic Circle. *Journal of Cleaner Production*, 194, pp. 473–482. https://doi.org/10.1016/j.jclepro.2018.05.125

Zeng, S. X., Liu, H. C., Tam, C. M., and Shao, Y. K. (2008). Cluster analysis for studying industrial sustainability: An empirical study in Shanghai. *Journal of Cleaner Production*, 16(10), pp. 1090–1097. https://doi.org/10.1016/j.jclepro.2007.06.004

11 An AHP-Based Approach to Determine the Effects of COVID-19 on Industrial Sustainability

Rohit Singh[1] *and Shwetank Avikal*[2]

[1]Systems Engineer, Infosys Limited, Bhubaneswar, India
[2]Department of Mechanical Engineering, Graphic Era Hill University, Dehradun, India

11.1 INTRODUCTION

The novel coronavirus initially started in Wuhan, China. Afterwards, it spread throughout the world. On January 30, 2020, India became part of this global carnage with the first COVID diagnosis (MOHFW). On March 11, 2020, the World Health Organisation (WHO) declared the outbreak of an influenza-like illness caused by the novel coronavirus SARS-CoV-2 (COVID-19) a pandemic. At present, with no proven effective treatment and vaccines, COVID-19 has continued to become a human and economic tragedy affecting millions of people.

To control the spread of the coronavirus or to "flatten the curve" of the infection, most governments around the world have put tight restrictions on the movement of people. This leads to a large-scale shut down of societies, factories, offices, colleges, non-essential shops; restricted traffic movement; and closing borders and restriction of mobility in public places.

The lockdown strategy comes with a heavy socio-economic price, particularly in low- and middle-income countries with a limited capacity to absorb prolonged national lockdowns (Petersen et al., 2020). According to the World Trade Organization, world trade is expected to fall by between 13% and 32% in 2020 as the COVID 19 pandemic disrupts normal economic activity and life around the world (World Trade Organization, 2020). It is estimated by WTO that the recovery in 2021 is equally uncertain, with outcomes depending largely on the duration of the outbreak and the effectiveness of the policy responses.

In the present work, the authors try to find out the impact of the outbreak on industries. These impacts have been studied in this study, and an AHP based approach has been used to find out the most influential and least influential impacts.

DOI: 10.1201/9781003102304-11

TABLE 11.1

Main Criteria and Their Sub-Criteria

S. No.	Criteria	Sub-Criteria
1.	Supply Chain Disruptions C1	Employees Layoff C11, Reduction in Production C12
2.	Financial C2	Impacts on Mortgages C21, Personal and commercial loans C22, Bankruptcy C23
3.	Transportation C3	Travel and Tourism C31, Transportation sector C32, Public Transport C33

11.2   PROBLEM DEFINITION

COVID-19 has already put several countries around the world in recession, especially low- and lower-middle-income countries. Many companies around the world filed for bankruptcy. The COVID-19 lockdown has broken the back of many big corporate firms, from oil to retail companies (Eggers, 2020). In the present study, we try to assess the impact of COVID-19 in industries and try to find out which criterion is most influential and which is least.

This problem is a multi-criteria decision-making problem that depends on a number of criteria. In the present work, three main criteria along with their sub-criteria have been considered. These criteria and sub-criteria are shown in Table 11.1. The description of each criterion and sub-criterion is in Table 11.2. For solving the discussed problem, a well-known decision-making approach Analytical Hierarchy Approach (AHP) has been used and has been discussed in the next section.

11.3   PROPOSED APPROACH

11.3.1   ANALYTICAL HIERARCHY PROCESS (AHP)

AHP is one of the most widely used approaches for solving multi-criteria decision-making problems. The concept of AHP was presented by Saaty in 1980. AHP leads with the measurement of pairwise comparisons and depends upon knowledge experts. In AHP, the decision-making problem is broken down into a simpler problem by making the comparisons of criteria using a scale of absolute judgment that represents how much one element dominates another with respect to a given attribute. Over the years, AHP has been widely used in many sectors. Baswaraj et al. (2018) have applied AHP for process parameter selection and consistency verification in secondary steel manufacturing. Ozdemir and Sahin (2018) have used AHP to evaluate locations taking into consideration both quantitative and qualitative factors in location selection for a Solar PV power plant. Kokangul et al. (2017) have applied AHP to prioritize the category of hazards based on experience.

TABLE 11.2
Description of Criteria and Sub-Criteria

Supply Chain Disruptions (C1)	COVID-19 have shown vulnerability in the current system of supply chain. It may finally forces many companies to transform their global supply chain.
Employees Layoff (C11)	Due to disruptions in supply chain many companies are forced to lay off their employee for survival on market.
Reduction in Production (C12)	The shutdown in China has prohibited import of various materials affecting both Indian manufacturers and component industry.
Financial (C2)	COVID-19 put a stain on financial situation. Small industries are in risk of run out of cash before reopening.
Impacts on Mortgages (C21)	The Mortgages industry is in severe stress. The mortgages are going to increase as the unemployment rises. Many households are unable to keep up with their mortgages.
Personal and commercial loans (C22)	The personal and Commercial loans are keep up piling on customer because of layoffs and reduction in salary.
Bankruptcy (C23)	COVID became the graveyard of many companies. The COVID-19 lockdown has break the back of many big corporate firms from oil companies to retail segment.
Transportation (C3)	Transport is very important in our daily life. But nowadays it is one of the most important issues that is to be considered. Transport of goods helps society but the transport of the public is dangerous these days.
Travel and Tourism (C31)	Travel and Tourism industry is among the most affected sector with massive fall in international demand
Transportation sector (C32)	It is most important to restart the transport of good throughout the country.
Public Transport (C33)	Public transport should not be reopened in the country. In includes railways, state transport busses, private busses, and local transport. It can support the public gathering and maintaining social distancing may be difficult for the public in the public transportation system.

11.4   RESULTS AND DISCUSSION

A decision-making approach, AHP, has been applied for evaluating the priority of each criterion and sub-criteria. For evaluating the priority of each criterion, each criterion has been compared with another one. A pairwise comparison matrix for the main criteria has been shown in Table 11.3.

The weights of these criteria have been calculated by AHP, and the results have been shown in Table 11.3. The results show that criteria C1 has the highest weight and criteria C2 and C3 have the next highest weight and the lowest weight, respectively (Table 11.4).

TABLE 11.3

Pairwise Comparison of Main Criteria

	C1	C2	C3
C1	1	2	2
C2	0.5	1	2
C3	0.5	0.5	1

TABLE 11.4

Main Criteria Weights Calculated by AHP

Criteria	Weight	CI & CR
C1	0.490476	CI = 0.026871CR = 0.051675
C2	0.311905	
C3	0.197619	

The sub-criteria of Supply Chain Disruption have also been compared with each other and they got equal importance. Their comparison matrix and weights have been shown in Table 11.5.

Similarly, the sub-criteria of Financial have also been compared with each other and a pairwise comparison matrix has been constructed and shown in Table 11.6. The comparative importance has been evaluated by AHP, and results have been shown in Table 11.6.

The sub-criteria of Transportation has also been compared pairwise and the comparison matrix and relative weights of each sub-criteria have been shown in Table 11.7.

The cumulative weights of each sub-criteria are calculated with the weights of each criterion and weights of each sub-criterion. The priority has been assigned to

TABLE 11.5

Pairwise Comparison Matrix and Weight of Sub-Criteria of Criteria Industries

	C11	C12	weights
C11	1	1	.50
C12	1	1	.50

TABLE 11.6

Pairwise Comparison Matrix and Weight of Sub-Criteria of Criteria Stores

	C21	C22	C23	Weights	CI & CR
C21	1	5	3	0.642196	CI = 0.021201CR = 0.04077
C22	0.2	1	0.5	0.121032	
C23	0.375	2	1	0.236772	

TABLE 11.7

Pairwise Comparison Matrix and Weight of Sub-Criteria of Criteria Transport

	C21	C22	C23	Weights	CI & CR
C21	1	2	5	0.594888	CI = 0.002769CR = 0.005326
C22	0.5	1	2	0.276611	
C23	0.2	0.5	1	0.128501	

TABLE 11.8

Cumulative Weights and Priority of Criteria and Sub-Criteria

Criteria/Sub-Criteria	Name	Weights	Cumulative Weights	Priority
C1	Supply Chain Disruption	0.490476		A
C11	Employees Layoff	.50	0.245238	1
C12	Reduction in Production	.50	0.245238	1
C2	Financial	0.311905		B
C21	Impacts on Mortgages	0.642196	0.200304	2
C22	Personal and commercial loans	0.121032	0.03775	6
C23	Bankruptcy	0.236772	0.07385	4
C3	Transportation C3	0.197619		C
C31	Travel and Tourism	0.594888	0.117561	3
C32	Transportation sector	0.276611	0.054664	5
C33	Public Transport	0.128501	0.025394	7

each sub-criterion on the basis of their cumulative importance and shown in Table 11.8. The priority of the main criteria has been assigned in terms of alphabets and in terms of numerical values for each sub-criterion and shown in Table 11.8.

11.5 CONCLUSION

India is facing the global problem of COVID-19. During the lockdown, several restrictions based on the concept of social distancing have been imposed on society. All the industries including, manufacturing, services sectors, transportation system and other essential services including government offices have been closed for minimizing the rate of the virus spread. The present study accesses the impact of COVID-19 in industrial sustainability. The results indicate that supply chain disruptions are the most influential impact faced by industries and public transport is the least impactful parameter.

COVID-19 have created new challenges for industries. It has shown the vulnerability of the industrial supply chain. In future work, many other factors could also be considered for the study and many other decision-making approaches could be implemented for solving this complex decision-making problem.

REFERENCES

Baswaraj, A., Rao, M.S., Pawar, P.J. (2018) Application of AHP for process parameter selection and consistency verification in secondary steel manufacturing. *Materials Today: Proceedings*, 5, pp. 27166–27170.

Eggers, F. (2020) Masters of disasters? Challenges and opportunities for SMEs in times of crisis. *Journal of Business Research*, 116, pp. 199–208.

https://www.mohfw.gov.in.

Kokangul, A., Polat, U., Suyu, C.D. (2017) A new approximation for risk assessment using the AHP and Fine Kinney methodologies. *Safety Science*, 91, pp. 24–32.

Ozdemir, S., Sahin, G. (2018) Multi-criteria decision-making in the location selection for a Solar PV power plant using AHP. *Measurement*, 129, pp. 218–226.

Petersen, E., Wasserman, S., Lee, S.S. et al. (2020) COVID-19–We urgently need to start developing an exit strategy. *International Journal of Infectious Diseases*, 96, pp. 233–239.

World Health Organisation (WHO) (2020) Rolling updates on coronavirus disease (COVID-19), https://www.who.int/emergencies/diseases/novel-coronavirus-2019/events-as-they.happen, Accessed 13 Apr.

WTO (2020) World Trade Organization, WTO. Trade set to plunge as COVID-19 pandemic upends global economy. https://www.wto.org/english/news_e/pres20_e/pr855_e.htm. 8 April 2020.

12 Roadmap to Smart Manufacturing for Developing Countries

Vaibhav S. Narwane[1], Rakesh D. Raut[2], and Balkrishna Eknath Narkhede[3]

[1]Department of Mechanical Engineering, K. J. Somaiya College of Engineering, Mumbai, India
[2]Department of Operations and Supply Chain Management, National Institute of Industrial Engineering (NITIE), Mumbai, India
[3]Department of Industrial Engineering, National Institute of Industrial Engineering (NITIE), Mumbai, India

12.1 INTRODUCTION

Industry 4.0 (I40) aims to create intelligent factories where technologies are transformed and upgraded by the means of Internet of Things (IoT), cyber-physical systems (CPSs), big data analytics, and cloud computing (Lu, 2017; Lasi et al., 2014). These "Digital twins" of the physical world and smart devices through sensors benefit manufacturing in terms of product customization, productivity, quality, visualization, sustainability, etc. (Rüßmann et al., 2015). Developed and emerging countries have created programs such as USA (Advanced Manufacturing Partnership), Germany (Industry 4.0), China (Made in China 2025), Korea (Innovation 3.0), Japan (Industry 4.1J), and Taiwan (Productivity 4.0).

However, developing countries follow a different structural transformation path than developed economies (Bah, 2011). Adoption of I40 is more challenging in developing countries due to poor digital infrastructure, imperfect standards, labour, etc. Dalenogare et al., (2018). Therefore, diffusion of Industry- 4.0 is slow in emerging economies, such as India, Brazil, and South Africa. Challenges to I40 in developing economy can be categorized as organizational, technological, strategic, legal, and ethical (Luthra and Mangla, 2018). Kamble et al. (2018) proposed four clusters of barriers for I40 adoption in manufacturing firms of India. Twelve barriers identified for the study were distributed into four clusters: autonomous, dependent, linkage, and drivers.

Lack of universal standards and protocol, absence of government policies and support, economic constraints, internet and infrastructure issues, high investment cost, and top management support are major barriers for I40 adoption in India

(Luthra and Mangla, 2018; Moktadir et al., 2018). Government support, in terms of subsidiaries to manufacturing firms, is a must in underdeveloped and developing economies (Szalavetz, 2018; Moktadir et al., 2018). With digitization, there are considerable changes in skill levels for job employment (Weber, 2016). Human resources is a major concern and must be addressed through training and organizational support (Agostini and Filippini, 2019).

Even though initially, I40 targets the manufacturing sector, its concept and vision can be extended to other sectors such as agriculture, service, food, healthcare, etc. Vertical, horizontal, and system integration offers an all-inclusive view of knowledge, tools, skills, and technologies for system autonomy: horizontal integration through service, planning, and procurement management; vertical integration through product, process, equipment, organization, and people (Pérez-Lara et al., 2019). This integration helps to improve the performance of service sector organizations. According to Geissbauer et al. (2016), the benefits of I40 are more in developing nations of the Asia-Pacific region. Advancements in mobile cloud computing and IoT have made an impact on the healthcare sector in terms of patient convenience, cost, and service availability (Roy et al., 2019). Improvements in big data for cyber-physical system (CPS) improve system efficiency, security, and scalability (Xu and Duan, 2019).

The purpose of this paper is to address the following research questions.

RQ1- What are critical issues of I40 adoption for developing countries?
RQ2- What are the applications of I40 other than the manufacturing sector?
RQ3- What smart manufacturing framework the organizations of developing countries need to adopt?

The structure of the paper is as follows: section 12.2 provides a literature survey on initiatives and technologies for I40; along with the adoption of I40. Section 12.3 presents the proposed framework. Section 12.4 concludes the study.

12.2   LITERATURE SURVEY

The literature review can be broadly categorized into three sections: (1) Initiatives taken towards Industry 4.0; (2) Technologies supporting Industry 4.0; (3) Adoption studies of Industry 4.0, followed by a research gap.

12.2.1   INITIATIVES TAKEN TOWARDS INDUSTRY 4.0

The 18th century was about the steam engine; the 19th century was about mass production with the help of assembly line; the 20th century included the new hardware and software required for manufacturing. A growing number of ICTs provide more knowledge to expand a manufacturing unit. In recent years, many initiatives have been taken by academicians as well as professionals in this domain. Most countries are building programs to address the market demand for smart manufacturing. Initiatives taken towards I40 are shown in Figure 12.1 and tabulated in Table 12.1.

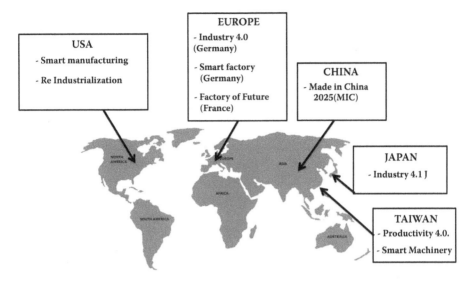

FIGURE 12.1 Initiative towards smart manufacturing/Industry 4.0.

TABLE 12.1
Initiatives Taken Towards I4.0

Sr. No.	Term	Founder	Year	Country	Reference
1	RepRap project	Dr Adrian Bowyer	2005	England	Jones et al. (2011)
2	Industry 4.0	German Gov.	2011	Germany	Roblek et al. (2016)
3	Smart factory	German Gov.	2012	Germany	Thoben et al. (2017)
4	Industrial Internet of Things (IIOT)	General Electric	2012	USA	Da Xu et al. (2014)
5	Smart manufacturing (SM)	USA Gov.	2012	USA	Davis et al. (2012)
6	Social Manufacturing	Prof. Feiyue Wang	2012	China	Shang et al. (2013)
7	Ford Freeform Fabrication technology (FFF)	Ford	2013	USA	Paniti and Somló (2014)
8	Future of Manufacturing (FoM)	UK Gov.	2013	UK	Chang et al. (2013)
9	Factories of the Future (FoF)	The European Commission	2014	Europe	Wang et al. (2016)
10	Innovation in Manufacturing 3.0	South Korea Gov.	2014	South Korea	Kang et al. (2016)
11	Made-in-China 2025 (MIC)	Chinese Gov.	2015	China	Wübbeke et al. (2016), Kennedy (2015)
12	Zero Downtime System (ZDT)	Fanuc Inc.	2016	Japan	Bogue (2017)
13	Industry 3.5	Chen-FuChien	2016	China	Chien et al. (2017)
14	Society 5.0	Japanese Gov.	2016	Japan	Esmaeilian et al. (2016)

RepRap stands for "Replicating Rapid prototype", which is an initiative developed to produce a low-cost 3D printer that can print most of its own components. The fourth industrial revolution will be set apart by full mechanization and digitization, and utilization of hardware and data advancements in manufacturing (Roblek et al., 2016). Inside the setting of the fourth industrial revolution, the term "Smart Factory" portrays a situation where machines and equipment can enhance forms through computerization and self-optimization. IIoT is a sub-paradigm of IoT, which is mostly focused on the interconnectivity of industrial assets like manufacturing tools and machines and logistic operations (Xu and Duan, 2019). The US government launched a smart manufacturing platform called "Smart Manufacturing Leadership Coalition". Social Manufacturing is a paradigm that allows customers to co-create products and services that are fully customized and personalized (Jiang et al., 2016). Ford's Freeform Fabrication technology (F3T) is a progressive system for sheet metal forming. The future of manufacturing initiated by the UK Government in 2013 looked at the long-term development of the manufacturing unit up to 2050 by targeting specific stages in the manufacturing value chain. "Factories of the Future" focuses on increasing the technologies used in the EU manufacturing sector.

Manufacturing Innovation 3.0 is a Korean government program established for the development of the manufacturing sector. Made in China 2025 is basically a vision set up by the Chinese Government to improve the manufacturing sector of China. Their prime focus was "quality over quantity". Productivity 4.0 initiated by Taiwan is about flexibility in the manufacturing sector. Zero Downtime system (ZDT) was designed by the company to completely eliminate downtime and increase overall efficiency. Industry 3.5 includes digital decision, smart manufacturing, supply chain, smart factory, and total resource management. The cores of Industry 4.0 and Industry 3.5 are completely different from each other. CPS is a core methodology of Industry 4.0 whereas TWD–Talents with Disabilities is a core methodology of Industry 3.5. In Society 5.0, a tremendous measure of data from sensors in physical space is collected on the internet.

12.2.2 Technologies Supporting Industry 4.0

In today's world, everyone is trying to implement the industrial revolution using advanced cutting edge information and communication technologies like the Internet of Things (IoT), Cloud computing, augmented reality, cyber-physical system (CPS) etc. in order to gain a competitive advantage. Table 12.2 shows the technologies supporting Industry 4.0.

The role of IoT in the smart factory is to enable accurate sensing and operational work, to provide accurate data for the interpretation of the results. Development in IoT in the following application domains is noticeable: the IoAT (Internet of Autonomous Things), the IoMT (Internet of Mobile Things), and the IoRT (Internet of Robotic Things). The smart factory has virtual resources and real-time accessibility of these resources. These resources are connected through networks (physical/hardware layer, cloud layer, consumer/user layer) and data can be stored over the

TABLE 12.2
Technologies Supporting I4.0

Sr. No.	Term	Founder	Year	Country	Reference
1	IoT	Kevin Ashton	1999	UK	Ashton (2009)
2	Cyber-physical system	NSF	2006	USA	Rajkumar et al. (2010)
3	Cloud computing	Google COE Eric Schmidt	2006	USA	Gartner (2009)
4	Smart grid	Energy Independence and security act	2007	USA	Federal Energy Regulatory Commission (2008)
5	Cloud manufacturing	Prof. Bo Hu Li	2009	China	Li et al. (2010)
6	Smart Augmented Reality tool	Airbus company	2010	France	Fite-Georgel (2011)
7	Cloud robotics	James J. Kuffner	2010	USA	Kehoe et al. (2015)
8	Cognitive computing	IBM	2011	USA	Kelly (2015)
9	CBDM	Dazhong Wu, David Rosen, and Dirk Schaefer	2012	USA	Wu et al. (2013)
10	Cloud-based remanufacturing	Wang X. V.	2014	China	Wang and Wang (2015)
11	Internet of Manufacturing Things	Zhang Y.	2014	China	Zhang et al. (2015)
12	A Cloud of Things (CoT)	Deutsche Telekom	2014	Germany	Aazam et al. (2014)
13	Internet Plus	Chinese Gov.	2015	China	Wang and Lin (2017)

internet using cloud computing. Integration of cloud computing and IoT leads to the formation of a concept called the cloud of things (CoT).

CoT is more knowledge-driven than data-driven. Every enterprise deals with complexity in decision-making activities, and CoT can be beneficial in marketing, logistics, supplier, manufacturer, and design data (Wang et al., 2014). In cloud manufacturing, all the manufacturing resources are sensed and connected to the cloud. For the sharing and exchanging the data automatically, IoT technologies like RFID tags and barcodes can be used (Zhong et al. 2017). Cloud-based Design and Manufacturing (CBDM) is a service-based manufacturing model in which end users can configure products or reconfigure the manufacturing system through the business models (Wu et al., 2013). In cloud-based remanufacturing, various information techniques are used in decision making for the future lifecycle of the product in the remanufacturing process (Esmaeilian et al., 2016). Cloud robotics is a field of application autonomy that endeavours to invoke cloud innovations, for example, Google's self-driving car which uses maps or images collected by the satellite, street views from the cloud to facilitate correct localization. Another example of cloud robotics is the Kiva Systems pallet robot for warehouse logistics (Kehoe et al., 2015).

CPS is a technology applicable to industries that integrates computational domains and physical processing to carry out a certain process. CPS constitutes a disruptive technique that incorporates automating regular, basic operations – taking into account Industry 4.0 – and working to insert intelligence in work processes to provide dynamic flexibility to the process to obtain high-quality goods at relatively lower prices. The Internet of Manufacturing Things is used to collect real-time data from shop floors with data capturing and processing, sensors and networks, etc. Internet Plus will play a very critical role in the next industrial revolution with the development of cell phones, cloud services, and big data analytics. The term "smart grid" refers to a new form of electricity framework including innovation, arrangement, and plans of action that are underway globally. Smart Augmented Reality Tool has been introduced with around a hundred tablets and been worked by in excess of 1000 customers at production offices across Europe. This framework is completely coordinated into the data framework and improves working circumstances and quality administration and control on the sequential assembly lines. Cognitive computing is the reproduction of human perspectives in an electronic model. Cognitive computing consists of the self-learning frameworks that use data mining, design acknowledgement and natural language to copy the way the human mind works.

12.2.3 ADOPTION OF INDUSTRY 4.0

Industry 4.0 is first adopted by developed countries, and then by developing and underdeveloped economics. I40 adoption developed countries has been studied by Basl (2017) in the Czech Republic, Jung et al. (2016) in the USA, Lin et al. (2018) in China, and Lu and Weng (2018) in Taiwan. To understand I40's penetration into ERP services and products, Basl (2017) surveyed fifteen ERP companies. Basl (2017) inferred that 73% of enterprises already have a strategy for I40 or in the process of making it.

Jung et al. (2016) proposed a plan of customized improvement based on current factory operations, readiness profile, and supply chain. IT maturity, regulatory policies, technological incentives, external pressure, and perceived benefits have a positive impact on I40; however, company size and nature does not have a positive impact (Lin et al., 2018). This shows that all type of manufacturing firms can benefit from smart manufacturing. Lu and Weng (2018) studied the impact of smart manufacturing in the electronic and computer product manufacturing industries of Taiwan. The role of BDA is important for identification/monitoring/warning and could be easily achieved through integrated machine equipment, tools and sensor devices.

Dachs et al. (2017), Lin et al. (2017), Schmidt et al. (2015), and Oesterreich and Teuteberg (2016) carried out I40 adoption studies in more than one country. To investigate whether smart manufacturing supports back-shoring, Dachs et al. (2017) conducted an empirical test of 2120 industrial firms from Switzerland, Germany, and Australia. The result shows that the readiness index of I40 is significant and its relationship with back-shoring is positive, and that also there is an increase in the flexibility of the manufacturing process. Lin et al. (2017)

investigated "China Manufacturing 2025" of China and the "Productivity 4.0 plan" of Taiwan from a government policy tool context. Data analysis show that weight priority for "political" is highest, i.e., China (49%) and Taiwan (44%). For China, the next top ranks are legal and regulatory (19%), public service (19%), and education (5%); while for Taiwan next top ranks are public services (25%), education (17%), and legal and regulatory (8%). Cross-country quantitative analysis of German-speaking countries was done by Schmidt et al. (2015). It hypothesised that three potential uses of I 4.0 are production time improvement, mass customization, and technology. Oesterreich and Teuteberg (2016) studied implications of automation and digitization for the construction industry using qualitative analysis. PESTEL analysis concluded that I40 for construction is beneficial from an economic and sustainability aspect, however, involvement of people, lack of protocol and standards, security, and fear of job loss are major concerns.

Developing countries like India, Singapore, and Brazil etc. have started the adoption of smart manufacturing. Ang et al. (2017), Kamble et al. (2018), Luthra and Mangla (2018), Tortorella and Fettermann (2018) carried out adoption studies in developing economies. Two different studies carried out by Kamble et al. (2018) and Luthra and Mangla (2018) investigated barriers and challenges to smart manufacturing for Indian companies. Lack of clear understanding about the benefits of IoT and higher cost of implementation are major barriers for Indian manufacturers (Kamble et al., 2018). Luthra and Mangla (2018) list the barriers as lack of universal standards and protocol, absence of government policies and support, economic constraints, internet and infrastructure issues, and top management support. Original equipment manufacturers (OEMs) like ship manufacturers face challenges of strict environmental rules and customized needs of the customer. Ang et al. (2017) proposed a 2-way framework for ship design, operation, and manufacturing with smart design (space exploration, performance evaluation, geometry modification), smart operation (big data), and smart manufacturing (Cyber-Physical Meta-Models, integration Meta-Models). Tortorella and Fettermann (2018) studied the relationship between lean practices and I40 implementation for manufacturing firms in Brazil. Lean practices and adoption levels of I4.0 were categorized into low-level and high-level and improvements in operational performance were studied for the same.

Szalavetz (2018) and Moktadir et al. (2018) conducted I40 adoption studies in Hungary and Bangladesh respectively, countries belonging to the semi or underdeveloped category. Government support in terms of subsidiaries manufacturing firms is must in these types of countries. Subsidies to advanced manufacturing technologies improve production capability, production activities, and technological activities; with subsidiary up-gradation and process development, R&D becomes intensive (Szalavetz, 2018). Moktadir et al. (2018) used a multiple case-study approach with four companies of leather processing. Results show that shortage of technological infrastructure, issues in reconfigurable manufacturing, integration of IT, and high investment cost were major concerns of manufacturers for the adoption of smart manufacturing.

Research Gaps

Literature survey shows that developed countries have already initiated I40, while developing and underdeveloped countries need to catch up with this new revolution. The challenges with I40 are technological, social, economic, and political. Thus the adoption of I40 for developing and developed economics is different. There is a need to understand implementation barriers in context with emerging economics. Also, the horizon of I4.0 needs to go beyond manufacturing and there is a need for this technology in sectors like health, service, agriculture, and food. However, these sectors should know the demerits and merits of I40 before its implementation.

12.3   PROPOSED FRAME WORK OF SMART MANUFACTURING

Figure 12.2 shows the proposed framework of the smart factory.

The adoption of smart manufacturing demands a holistic approach. The consideration of technological, social, economic, and political factors must be there. Smart manufacturing needs advanced infrastructure, considerable investment, and organizational support. Top management and training plays an important role in adoption. Smart manufacturing encourages the use of sustainable practices. Organizations of a developing country like India need the of the State as well as National Governments through policy-making and subsidy.

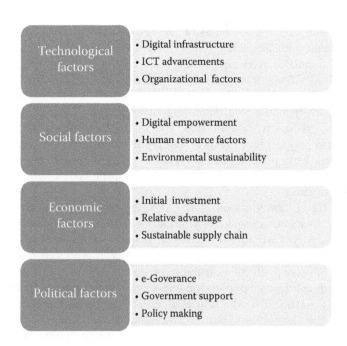

FIGURE 12.2   Proposed framework of smart manufacturing.

12.4 DISCUSSION

As mentioned in the literature survey section, different countries have taken initiatives towards I40 with different names. Even though Germany has the credit of being the originator of the concept, it spread around all parts of words, including the USA, Japan, China, France etc. I40 needs the support and initiative of the government; however industrial firms and consulting companies must take an active part in its success. The Association of Southeast Asian Nations wants to take advantage of I40 in data sharing and knowledge diffusion across Indonesia to Vietnam (ASEAN, 2018). "Make in India" and "Digital India" campaigns by India have had a positive impact on the development of the manufacturing sector (Karani and Panda, 2018). However, insufficient budget and lack of infrastructure are major barriers in the implementation of I40 in developing economies. Even after this initiative, individuals will continue to play an active, engaging role in manufacturing.

Smart manufacturing systems associated with IoT is described as the "intelligent connectivity of smart devices by which objects can sense one another and communicate". The manufacturing equipment is connected to a Web-based network and must use open software architecture, open protocols, open data models for better connectivity. As devices are connected, the data they generate runs into software applications that create the information which individuals can use to make choices that are timely and effective. This improves the performance. However cybersecurity is a major issue, cyber threats to the system are real, global, and growing. Businesses should protect their data, systems, and networks at every step toward becoming part of the I40. Challenges can be categorized into architectural, business, hardware, privacy, standard, and technical.

The economic boost in industries due to the use of I40 is considerable. The key issues that need to be addressed are energy efficiency and autonomy, open networks, and location-independent human-readable identification, and device-independent application development facilities.

REFERENCES

Aazam, M., Khan, I., Alsaffar, A. A., and Huh, E. N. (2014, January). Cloud of Things: Integrating Internet of Things and cloud computing and the issues involved. In *2014 11th International Bhurban Conference on Applied Sciences and Technology (IBCAST)* (pp. 414–419). IEEE.

Agostini, L., and Filippini, R. (2019). Organizational and managerial challenges in the path towards Industry 4.0. *European Journal of Innovation Management*, 22(3), pp. 406–421.

Ang, J., Goh, C., Saldivar, A., and Li, Y. (2017). Energy-efficient through-life smart design, manufacturing and operation of ships in an industry 4.0 environment. *Energies*, 10(5), Art. no. 610.

ASEAN. (2018). *About ASEAN*. Retrieved from Association of Southeast Asian Nations: http://asean.org/asean/about-asean/ (accessed 06 February 2019)

Ashton, K. (2009). That 'internet of things' thing. *RFID journal*, 22(7), pp. 97–114.

Bah, E. H. M. (2011). Structural transformation paths across countries. *Emerging Markets Finance and Trade*, 47(sup2), pp. 5–19.

Basl, J. (2017). Pilot study of readiness of Czech companies to implement the principles of Industry 4.0. *Management and Production Engineering Review*, 8(2), pp. 3–8.

Bogue, R. (2017). Cloud robotics: A review of technologies, developments and applications. *Industrial Robot: An International Journal*, 44(1), pp. 1–5.

Chang, H. J., Andreoni, A., and Kuan, M. L. (2013). International industrial policy experiences and the lessons for the UK, 5–74.

Chien, C. F., Hong, T. Y., and Guo, H. Z. (2017). An empirical study for smart production for TFT-LCD to empower Industry 3.5. *Journal of the Chinese Institute of Engineers*, 40(7), pp. 552–561.

Da Xu, L., He, W., and Li, S. (2014). Internet of things in industries: A survey. *IEEE Transactions on Industrial informatics*, 10(4), pp. 2233–2243.

Dachs, B., Kinkel, S., and Jäger, A. (2017). Bringing it all back home? Backshoring of manufacturing activities and the adoption of Industry 4.0 technologies, 54(6), pp. 101017.

Dalenogare, L. S., Benitez, G. B., Ayala, N. F., and Frank, A. G. (2018). The expected contribution of Industry 4.0 technologies for industrial performance. *International Journal of Production Economics*, 204, pp. 383–394.

Davis, J., Edgar, T., Porter, J., Bernaden, J., and Sarli, M. (2012). Smart manufacturing, manufacturing intelligence and demand-dynamic performance. *Computers & Chemical Engineering*, 47, pp. 145–156.

Esmaeilian, B., Behdad, S., and Wang, B. (2016). The evolution and future of manufacturing: A review. *Journal of Manufacturing Systems*, 39, pp. 79–100.

Federal Energy Regulatory Commission (2008). Federal Energy Regulatory Commission Assessment of Demand Response & Advanced Metering, Staff Report and Excel Data, 29 December 2008.

Fite-Georgel, P. (2011, October). Is there a reality in industrial augmented reality? In *2011 10th IEEE International Symposium on Mixed and Augmented Reality (ISMAR)* (pp. 201–210). IEEE.

Geissbauer, R., Vedsø, J., and Schrauf, S. (2016). A strategist's guide to industry 4.0. *Strategy & Business*, 83, pp. 6–47.

Jiang, P., Ding, K., and Leng, J. (2016). Towards a cyber-physical-social-connected and service-oriented manufacturing paradigm: Social manufacturing. *Manufacturing Letters*, 7, pp. 15–21.

Jones, R., Haufe, P., Sells, E., Iravani, P., Olliver, V., Palmer, C., and Bowyer, A. (2011). RepRap–the replicating rapid prototyper. *Robotica*, 29(1), pp. 177–191.

Jung, K., Kulvatunyou, B., Choi, S., and Brundage, M. P. (2016, September). An overview of a smart manufacturing system readiness assessment. In *IFIP International Conference on Advances in Production Management Systems* (pp. 705–712). Springer, Cham.

Kamble, S. S., Gunasekaran, A., and Sharma, R. (2018). Analysis of the driving and dependence power of barriers to adopt industry 4.0 in Indian manufacturing industry. *Computers in Industry*, 101, pp. 107–119.

Kang, H. S., Lee, J. Y., Choi, S., Kim, H., Park, J. H., Son, J. Y., and Do Noh, S. (2016). Smart manufacturing: Past research, present findings, and future directions. *International Journal of Precision Engineering and Manufacturing-Green Technology*, 3(1), pp. 111–128.

Karani, A., and Panda, R. (2018). 'Make in India' campaign: Labour law reform strategy and its impact on job creation opportunities in India. *Management and Labour Studies*, 43(1-2), pp. 58–69.

Kehoe, B., Patil, S., Abbeel, P., and Goldberg, K. (2015). A survey of research on cloud robotics and automation. *IEEE Transactions on Automation Science and Engineering*, 12(2), pp. 398–409.

Gartner (2009). Cloud computing inquiries at Gartner. Available at: http://blogs.gartner.com/thomas_bittman/2009/10/29/cloud-computing-inquiries-at-gartner (accessed 16 December 2018).

Kelly, J. E. (2015). Computing, cognition and the future of knowing, White Paper, IBM Research, 2.

Kennedy, S. (2015). Made in China 2025. Center for Strategic and International Studies.

Lasi, H., Fettke, P., Kemper, H. G., Feld, T., and Hoffmann, M. (2014). Industry 4.0. *Business & Information Systems Engineering*, 6(4), pp. 239–242.

Li, B. H., Zhang, L., Wang, S. L., Tao, F., Cao, J. W., Jiang, X. D., and Chai, X. D. (2010). Cloud manufacturing: A new service-oriented networked manufacturing model. *Computer integrated manufacturing systems*, 16(1), pp. 1–7.

Lin, K. C., Shyu, J. Z., and Ding, K. (2017). A cross-strait comparison of innovation policy under Industry 4.0 and sustainability development transition. *Sustainability*, 9(5), Art. no. 786.

Lin, D., Lee, C. K. M., Lau, H., and Yang, Y. (2018). Strategic response to Industry 4.0: An empirical investigation on the Chinese automotive industry. *Industrial Management & Data Systems*, 118(3), pp. 589–605.

Lu, Y. (2017). Industry 4.0: A survey on technologies, applications and open research issues. *Journal of Industrial Information Integration*, 6, pp. 1–10.

Lu, H. P., and Weng, C. I. (2018). Smart manufacturing technology, market maturity analysis and technology roadmap in the computer and electronic product manufacturing industry. *Technological Forecasting and Social Change*, 133, pp. 85–94.

Luthra, S., and Mangla, S. K. (2018). Evaluating challenges to Industry 4.0 initiatives for supply chain sustainability in emerging economies. *Process Safety and Environmental Protection*, 117, pp. 168–179.

Moktadir, M. A., Ali, S. M., Kusi-Sarpong, S., and Shaikh, M. A. A. (2018). Assessing challenges for implementing Industry 4.0: Implications for process safety and environmental protection. *Process Safety and Environmental Protection*, 117, pp. 730–741.

Oesterreich, T. D., and Teuteberg, F. (2016). Understanding the implications of digitisation and automation in the context of Industry 4.0: A triangulation approach and elements of a research agenda for the construction industry. *Computers in Industry*, 83, pp. 121–139.

Paniti, I., and Somló, J. (2014). Novel incremental sheet forming system with tool-path calculation approach. *Acta Polytechnica Hungarica*, 11(7), pp. 43–60.

Pérez-Lara, M., Saucedo-Martínez, J. A., Salais-Fierro, T. E., Marmolejo-Saucedo, J. A., and Vasant, P. (2019). Vertical and horizontal integration systems in Industry 4.0. In *Innovative Computing Trends and Applications* (pp. 99–109). Springer, Cham.

Rajkumar, R., Lee, I., Sha, L., and Stankovic, J. (2010, June). Cyber-physical systems: The next computing revolution. In *2010 47th ACM/IEEE Design Automation Conference (DAC)* (pp. 731–736). IEEE

Roblek, V., Meško, M., and Krapež, A. (2016). A complex view of industry 4.0. *Sage Open*, 6(2), Art. no. 2158244016653987.

Roy, S., Das, A. K., Chatterjee, S., Kumar, N., Chattopadhyay, S., and Rodrigues, J. J. (2019). Provably secure fine-grained data access control over multiple cloud servers in mobile cloud computing based healthcare applications. *IEEE Transactions on Industrial Informatics*, 15(1), pp. 457–468.

Rüßmann, M., Lorenz, M., Gerbert, P., Waldner, M., Justus, J., Engel, P., and Harnisch, M. (2015). Industry 4.0: The future of productivity and growth in manufacturing industries. *Boston Consulting Group*, 9.

Schmidt, R., Möhring, M., Härting, R. C., Reichstein, C., Neumaier, P., and Jozinović, P. (2015, June). Industry 4.0-potentials for creating smart products: Empirical research results. In *International Conference on Business Information Systems* (pp. 16–27). Springer, Cham.

Shang, X., Liu, X., Xiong, G., Cheng, C., Ma, Y., and Nyberg, T. R. (2013, July). Social manufacturing cloud service platform for the mass customization in apparel industry. In *2013 IEEE International Conference on Service Operations and Logistics, and Informatics (SOLI)* (pp. 220–224). IEEE.

Szalavetz, A. (2018). Industry 4.0 and capability development in manufacturing subsidiaries. *Technological Forecasting and Social Change*, 145, pp. 384–395.

Thoben, K. D., Wiesner, S., and Wuest, T. (2017). "Industrie 4.0" and smart manufacturing–a review of research issues and application examples. *International Journal of Automotive Technology*, 11(1), pp. 4–16.

Tortorella, G. L., and Fettermann, D. (2018). Implementation of Industry 4.0 and lean production in Brazilian manufacturing companies. *International Journal of Production Research*, 56(8), pp. 2975–2987.

Wang, Y. J., and Lin, H. Y. (2017). Research on internet plus innovative diversified education mode. *DEStech Transactions on Social Science, Education and Human Science* (eiem), pp. 6–9.

Wang, X. V., and Wang, L. (2015). WRCloud: A novel WEEE remanufacturing cloud system. *Procedia CIRP*, 29, pp. 786–791.

Wang, L., Wang, X. V., Gao, L., and Váncza, J. (2014). A cloud-based approach for WEEE remanufacturing. *CIRP Annals-Manufacturing Technology*, 63(1), pp. 409–412

Wang, S., Wan, J., Li, D., and Zhang, C. (2016). Implementing smart factory of industrie 4.0: An outlook. *International Journal of Distributed Sensor Networks*, 12(1), Art. no. 3159805.

Weber, E. (2016). *Industry 4.0: Job-Producer or Employment-Destroyer?* (No. 2/2016). Aktuelle Berichte.

Wu, D., Thames, J. L., Rosen, D. W., and Schaefer, D. (2013). Enhancing the product realization process with cloud-based design and manufacturing systems. *Journal of Computing and Information Science in Engineering*, 13(4), pp. 041004-1–041004-14.

Wübbeke, J., Meissner, M., Zenglein, M. J., Ives, J., and Conrad, B. (2016). Made in China 2025. *Mercator Institute for China Studies*, 2, pp. 14–41.

Xu, L. D., and Duan, L. (2019). Big data for cyber physical systems in industry 4.0: A survey. *Enterprise Information Systems*, 13(2), pp. 148–169.

Zhang, Y., Zhang, G., Wang, J., Sun, S., Si, S., and Yang, T. (2015). Real-time information capturing and integration framework of the internet of manufacturing things. *International Journal of Computer Integrated Manufacturing*, 28(8), pp. 811–822.

Zhong, R. Y., Xu, X., Klotz, E., and Newman, S. T. (2017). Intelligent manufacturing in the context of industry 4.0: A review. *Engineering*, 3(5), pp. 616–630.

13 Sustainable Entrepreneurship

An Emerging Concept Benefiting Small and Medium Entrepreneurs and Their Future

Tanuja Gour[1] and Amar Singh[2]
[1]Assistant Professor, School of Management, IMS Unison University, Dehradun, 248009, India
[2]Assistant Professor, School of Commerce, Graphic Era Hill University, Dehradun, Uttarakhand, 248001, India

13.1 INTRODUCTION

Over years, numerous viewpoints to sustainability have developed, making it significant from both visionary and application standpoints. To thrive, progress, and find solutions, entrepreneurs make themselves familiar with society and the environment. They take initiatives towards the progress of the community and the rights of the citizens. Entrepreneurs also recognize that it is essential to ensure convenient functioning for all people, including their health, finances, and social aspects. This leads to a new prospect on the profitability of the business, currently called the performance of business where both social as well as environmental matters are core foundations of sustainability (Martínez-Ferrero et al., 2015). Thus, sustainable entrepreneurship is referred to as the combination of social aspects (investors like employees, business associates, suppliers, market, etc.), environmental issues (associated with a long-range shield with reducing harmful effect) as well as economic factors (economic growth) (Crowther & Aras, 2008).

13.2 THE CONCEPT OF SUSTAINABILITY

The term sustainability has its origin in Latin. It emanates from "*sustinere*", in which "*tinere*" means "to hold" and "*sus*" means "under". Sustainability is a means of creating an energetic environment and a high-quality life, as well as appreciating

DOI: 10.1201/9781003102304-13

the need to maintain and conserve the available resources of the world. It is based on the theory that future generations must breathe on a planet that the contemporary individual benefits from (Clough et al., 2006). Sustainability aims at ensuring a good life for everyone. It is about exploring a way of living that ensures a good life for everybody. Sustainability is associated with the discipline of ecology.

Traditionally, the word "sustainable" is based around environmental concerns, and almost all literature and analysis tools show this significance. Although, it is gradually noticed that sustainability cannot be attained if social issues are not considered. Weitzman (1997) suggested that sustainability is an estimate of prospective expenditure. Conway & Barbier (1990) proposed *sustainability of economic* concerns with ***the capability to retain productiveness.*** It is essential that to sustain any business for a longer period, the organization should not only concentrate on maximizing profit but should be attentive to the environment, and show their accountability towards the community (Mojekeh & Eze, 2011).

13.3 SUSTAINABLE ENTREPRENEURSHIP (SE)

Entrepreneurship and sustainability management have together led to the establishment of an advanced method known as "Sustainable Entrepreneurship" (Dean & McMullen, 2007). For Sustainable Entrepreneurship, business owners must comply with prime accountabilities towards their financial sponsors as well as stockholders, along with nature (environment), society, and future generations. Cohen and Winn (2007) described "Sustainable Entrepreneurship" from the perspective of universal advantage as the essentials of how "future" products and services are revealed, formed, and operated by whom and in what ways. Patzelt and Shepherd (2011) defined Sustainable Entrepreneurship as *"the discovery, creation, and exploitation of opportunities to create future goods and services that sustain the natural and communal environment and provide development gain for others"*.

SE is an emerging concept focussed on future innovations (Cohen & Winn, 2007), and the resolution of economic, social, and environmental issues. Sustainable entrepreneurs need to be specific about what needs to be maintained, such as ecology, life support, and families, and what needs to be developed, like economic well-being, along with social and cultural benefits (Patzelt & Shepherd, 2011). In such a scenario, new businesses will merge fiscal, social, and environmental problems and contribute to a different form of entrepreneurship called sustainable entrepreneurship (Hall et al., 2010). It has been recommended for those entrepreneurs who embody various sustainability values and aim at constructing an attitude to set benchmarks to be coined in the future (Leiserowitz et al., 2006). Schaltegger and Wagner (2011) described SE as "a creative, market-oriented and personality-driven philosophy of producing inventive goods during the initial stage of the business to support humanity or the world". However, Sustainable Entrepreneurship can lead to long-term economic, social, and environmental benefits. SE merges objectives of SD (Jacobs, 1995) in addition to entrepreneurial act with the growth of the economy (Gibbs, 2008). Kuckertz and Wagner (2010) suggested that SE includes all commercial activities that focus on SD and aim to benefit from them.

13.4   THREE PILLARS OF SUSTAINABLE ENTREPRENEURSHIP (TBL): PEOPLE, PROFIT, AND PLANET

It has been observed that sustainability refers to the dynamic equilibrium between its three major elements i.e., environment, social welfare of individuals, and economic actions. These three correlated and complementary elements are referred to as the "Triple Bottom Line" (TBL) of Sustainable Entrepreneurship. This was suggested by Freer Spreckley in the year 1981 (Spreckley, 1981) then described elaborately by John Elkington in 1997. He explained that once an organization introduces TBL, a business will not solely concentrate on economic issues, but also on societal and environmental issues that they sometimes include or exclude (Elkington, 1997). In 2004, the theory was amended to include the "3Ps formulation": People, Planet, and Profit (Elkington, 2004). "People" represents firms' responsibility towards society, "Planet" is the responsibility towards the environment, and "Profit" is the responsibility towards economic benefits.

The Triple Bottom Line (TBL) theory has been used popularly by many researchers for explaining the term "Sustainable Development" (Chick, 2009). Majid and Koe (2012) revised the model of Sustainable Entrepreneurship based on the Triple Bottom Line (TBL) theory and recommended that economic, social, ecological, and cultural attributes are important for sustainable entrepreneurship. According to Slaper & Hall, (2011), the adaptability of the TBL concept in the organizations must be implemented properly so that beyond the foundation of quantifying the sustainability on three bases i.e., People, Planet, and Profits, it can easily meet our firm's specific requirements. The organization's challenge involves calculating every category of 3Ps, search for applicable data, and measuring a project or plan to contribute towards sustainability.

The frame of the Triple Bottom Line demands firms to measure the consequences of their resolutions from a long-term panorama. According to (Batra, 2012),

- "People" defines how an enterprise treats its labour force.
- "Planet" defines the influence of enterprise on possessions of nature as well as on the environment; and
- "Profit" is related not only to the returns on investment also to allocate financial revenues among investment along with profit dispersal.

According to Racelis, (2014) the phrase "Triple Bottom Line" recommends three distinct (and separate) bottom lines: (i) the conventional measurement of corporate profit – the bottom line of the "Profit and loss account", (ii) the bottom line of the organization's "People account" – a measure in some form of how an organization has been socially responsible throughout its operations process, and (iii) the bottom line of the organization's "Planet account" – a measurement of how it has been responsible towards the environment. TBL aims is to measure the financial, social, and environmental aspects and performance of the enterprise over a period.

13.5   SUSTAINABLE ENTREPRENEURSHIP BENEFITTING SMES

In a developing country like India, according to MSMED (Micro, Small and Medium Enterprises Development) Act, 2006, enterprises that come under the micro, small, and medium category are bifurcated into two sections i.e., manufacturing and service enterprises. The firms involved in the manufacturing or production process of products are referred to as per the firms' investment in plant and machinery. The current study is focused only on small and medium scale industries.

Small-scale enterprises are those whose investment limit in plant and machinery must be more than Rs. 25 lacs and less than Rs.5 crores. On the other hand, medium enterprises are those whose investment in plant and machinery must exceed Rs. 5 crores but should be less than Rs. 10 crores. The World Bank outlines small enterprises as enterprises where the number of employees must be up to 50, total assets and sales must be up to US$ 3million, and medium enterprise where the number of employees is limited to 300 with total assets and sales up to US$ 15 million (Ayyagari et al., 2007).

Mitra and Pingali (1999) suggested that small and medium industries are the main employment-generating entities and have an advantage of a low-cost workforce, operational activities, and indigenous technical knowledge. Many studies have suggested that small firms have an essential role in employment generation and that small and medium entrepreneurs will continuously help sustain the progress and expansion in emerging countries (Smallbone & Wyer, 2000; Curran 2000; Davidsson & Delmar, 1997). Nowduri (2012) explained that small- and medium-scale firms are highly influential in financial expansion in a justifiable and impartial manner all over a country. Small-scale and medium firms require proper administration along with well-organized operational activities. Hence, accurate knowledge is essential to improve the efficiency of small business owners (DecSanctis, 2003).

The policies and predictions in a sustainable economy depend extensively on the manufacturing, technical, and administrative processes of small-scale as well as medium-scale industries (Biondi et al., 2002). It must be ensured that small and medium firms participate in the process of sustainable progress. Small and medium business owners face many difficulties such as financial constraints, lack of technical knowledge, absence of required information, organizational values, and inner motivation in applying sustainable prospects (Hemel & Cramer, 2002). The study revealed by researchers that small firms were far from applying sustainability in their business (Omar & Samuel, 2011). The concept of Sustainable Entrepreneurship is very new and emerging, and there is a lack of research in small business firms (Dixon & Clifford, 2007) in emerging economies (Rasi et al., 2010). Small business owners comprise a friendly environment for entrepreneurs (Badenhorst et al., 1998). Entrepreneurship is known as a significant asset in developing, stabilizing, and sustaining a rich heritage in the whole world (Phelps, 2007). However, the influence of small-scale industries towards economic growth and prosperity must not be ignored (Ukenna et al., 2010). As per Ojukwu's (2006) study, small firms and medium entrepreneurs offer the foundation through which economic prosperity and sustainability of any nations exist. Furthermore, SMEs promote entrepreneurship, generate employment opportunities for many people, use and organize investment and produce raw materials for large-scale manufacturing enterprises.

13.6 CONCLUSION

Any business organization achieves sustainability and stability only when it changes its perception towards economic, social, and environmental factors. Much research has been conducted to support and assist Sustainable Entrepreneurship that not only focuses on the protection and safety of the environment but must include economic constraints and social issues. Entrepreneurs who are capable of maintaining and balancing their efforts in all three areas i.e., economic, social, and environmental, are considered Sustainability entrepreneurs. The contributions of SMEs over the past many years cannot be ignored as they employ a large number of people, whether skilled, semi-skilled, or unskilled. The government also encourages SMEs so that more and more entrepreneurs can start their business and provide job opportunities as these are more labour-intensive. Despite many advantages, it is a bitter truth that SMEs face many obstacles including finance, less educational qualification, lack of technical knowledge, organizational culture, motivational spirit, innovative ideas, lack of awareness about government schemes and financial assistance, family and relatives' support and so on, which may discourage them to further expand their operations.

REFERENCES

Ayyagari, M., Beck, T., and Demirguc-Kunt, A. (2007). Small and medium enterprises across the globe. *Small Business Economics*, 29.4, pp. 415–434. Doi: 10.1007/s11187-006-9002-5.

Badenhorst, J. A., de J. Cronje, G. J., du Toit, G. S., Gerber, P. D., Kruger, L. P., de K. Marais, A., Strydom, J. W., van der Walt, A., and van Reneen, M. J. (1998). *Business Management* (4th Ed). South Africa: International Thomson Publishing.

Batra, S. (2012). Sustainable entrepreneurship and knowledge-based development. In *Proceedings of the Eleventh International Entrepreneurship Forum Kuala Lumpur, Malaysia* (pp. 3–6).

Biondi, V., Iraldo, F., and Meredith, S. (2002). Achieving sustainability through environmental innovation: The role of SMEs. *International Journal of Technology Management*, 24(5-6), pp. 612–626.

Clough, G. W., Chameau, J. L., and Carmichael, C. (2006). Sustainability and the university. *Presidency*, 9(1), p. 30–32, 35, 37–38. https://eric.ed.gov/?id=EJ796131

Chick, A. (2009). Green entrepreneurship: A sustainable development challenge. In Mellor, R., Coulton, G., Chick, A., Bifulco, A., Mellor, N., and Fisher, A. (Eds.), *Entrepreneurship for Everyone: A Student Textbook* (pp. 139–150). London: SAGE Publications.

Conway, G. R., and Barbier, E. B. (1990). *After the Green Revolution: Sustainable Agriculture for Development*. London: Earthscan Publications.

Cohen, B., and Winn, M.I. (2007). Market imperfections, opportunity and sustainable entrepreneurship. *Journal of Business Venturing*, 22(1), pp. 29–49.

Crowther, D., and Aras, G. (2008). *Corporate social responsibility*- eBooks and textbooks from bookboon.com

Curran, J. (2000). What is small business policy in the UK for? Evaluation and assessing small business policies. *International Small Business Journal*, 18(3), pp. 36–50.

Davidsson, P., and Delmar, F. (1997). High-growth firms: characteristics, job contribution and method observations. In RENT XI Conference.

Dean, T. J., and McMullen, J. S. (2007). Toward a theory of sustainable entrepreneurship: Reducing environmental degradation through entrepreneurial action. *Journal of Business Venturing*, 22(1), pp. 50–76

DecSanctis, G. (2003). The social life information systems research. *Journal of Association of Information Research*, 4(6), pp. 360–376.

Dixon, S.E., and Clifford, A. (2007). Ecopreneurship: A new approach to managing the triple bottom line. *Journal of Organizational Change Management*, 20(3), pp. 326–345.

Elkington, J. (2004). Enter the triple bottom line. In Adrian Henriques and Julie Richardson (Eds), *The Triple Bottom Line: Does It All Add Up*. EarthScan: London UK.

Elkington, J. (1997). *Cannibals with Forks: the Triple Bottom Line of 21st Century Business*. Oxford: Capstone Publishing Ltd.

Gibbs, D. (2008). Sustainability entrepreneurs, ecopreneurs and the development of a sustainable economy. *Greener Management International*, 55, pp. 63–78.

Hall, J.K., Daneke, G.A., and Lenox, M.J. (2010). Sustainable development and entrepreneurship: Past contributions and future directions. *Journal of Business Venturing*, 25, pp. 439–448.

Hemel, C., and Cramer, J. (2002). Barriers and stimuli for eco design in SMEs. *Journal of Cleaner Production*, 10, pp. 439–453.

Jacobs, M. (1995). Sustainable development, capital substitution and economic humility: A response to Beckerman. *Environmental Values*, 4 (1), pp. 57–68.

Kuckertz, A., and Wagner, M. (2010). The influence of sustainability orientation on entrepreneurial intentions—Investigating the role of business experience. *Journal of Business Venturing*, 25(5), pp. 524–539.

Leiserowitz, A. A., Kates, R. W., and Parris, T.M. (2006). Sustainability values, attitudes, and behaviors: A review of multinational and global trends. *Annual Review of Environment and Resources*, 31(1), pp. 413–444.

Majid, I. A. and Koe, W.L. (2012). Sustainable entrepreneurship (SE): A revised model based on triple bottom line (TBL). *International Journal of Academic Research in Business and Social Sciences*, 2(6), pp. 293–310, ISSN: 2222-6990.

Martịnez-Ferrero, J., Garcia-Sanchez, I. M., and Cuadrado-Ballesteros, B. (2015). Effect of financial reporting quality on sustainability information disclosure. *Corporate Social Responsibility and Environmental Management*, 22(1), pp. 45–64.

Mitra, R., and Pingali, V. (1999). Analysis of growth stages in small firms: A case study of automobile ancillaries in India. *Journal of Small Business Management*, 37(3), pp. 62–75.

Mojekeh, M. O., and Eze, P. A. (2011). The environmental impact of production and sales of sachet water in Nigeria. *African Research Review*, 5(4), pp. 479–492.

Nowduri, S. (2012). Framework for sustainability entrepreneurship for small and medium enterprises (SMEs) in an emerging economy, *World Journal of Management*, 4(1), pp. 51–66.

Omar, R., and Samuel, R. (2011). Environmental management amongst manufacturing firms in Malaysia. In *2011 3rd International Symposium and Exhibition in Sustainable Energy and Environment (ISESEE)* (pp. 148–151). IEEE.

Ojukwu, D. (2006). Achieving sustainable growth through the adoption of integrated business of Nigerian small and medium sized enterprises. *Journal of Information Technology Impact*, 6(1), pp. 41–60.

Patzelt, H., and Shepherd, D. A. (2011). Recognizing opportunities for sustainable development. *Entrepreneurship Theory and Practice*, 35(4), pp. 631–652.

Phelps, E. S. (2007). Entrepreneurial culture. *The Wall Street Journal*, 12.

Racelis, A. D. (2014). Sustainable entrepreneurship in Asia: A proposed theoretical framework based on literature review. *Journal of Management for Global Sustainability*, 2(1), pp. 49–72.

Rasi, R. R. M., Abdekhodaee, A., and Nagarajah, R. (2010). Understanding drivers for environmental practices in SMEs: A critical review. In *2010 IEEE International Conference on Management of Innovation and Technology (ICMIT)* (pp. 372–377). IEEE.

Schaltegger, S., and Wagner, M. (2011). Sustainable entrepreneurship and sustainability innovation: Categories and interactions. *Business Strategy and the Environment*, 20(4), pp. 222–237.

Slaper, T. F., and Hall, T. J. (2011). The triple bottom line: What is it and how does it work? *Indiana Business Review*, 86(1), pp. 4–8.

Smallbone, D., and Wyer, P. (2000). Growth and development in the small firm. In S. Carter and D. Jones-Evans (Eds), *Enterprise and Small Business: Principles, Practice and Policy*. Harlow: Financial Times Prentice Hall, pp. 409–433.

Spreckley, F. (1981). *Social Audit - A Management Tool for Co-operative Working*. Leeds: Beechwood College.

Tilley, F., and Young, W. (2009). Sustainability entrepreneurs: Could they be the true wealth generators of the future? *Green Management International*, 55, pp. 79–92.

Ukenna S., Ijeoma N., Anionwu C., and Olise M. C. (2010). Effect of investment in human capital development on organizational performance: Empirical examination of the perception of small business owners in Nigeria. *European Journal of Economics, Finance and Administrative Sciences*, 26, pp. 93–107.

Weitzman, M. L. (1997). Sustainability and technical progress. *Scandinavian Journal of Economics*, 99(1), pp. 1–13. https://onlinelibrary.wiley.com/doi/pdf/10.1111/1467-9442.00043.

Index